Advances in Military Geosciences

Advances in Military Geosciences is a book series which explores the interaction between current and historic military operations and earth science, including geography, geology, geophysics, soil science, ecology, hydrology, glaciology and atmospheric sciences.

Military activities are almost always strongly integrated within a wide spectrum of geoscience. The decisive outcomes of land battles throughout history have been dictated in large part by the terrain and environmental setting. Modern military operations rely on a wide range of land-, air-, sea-, and space-borne intelligence and knowledge of dynamic terrain processes and conditions. In addition, the study of geo-based environmental science is critical to both the sustainable management of military reservations and installations, as well as the evaluation of how terrain and environmental conditions may impact military equipment and operations.

Advances in Military Geosciences contains single and multi-authored books as well as edited volumes. Series Editors are currently accepting proposals, forms for which can be obtained from the publisher, Ron Doering (ron.doering@springer.com).

More information about this series at http://www.springer.com/series/15030

Peter L. Guth
Editor

Military Geoscience

Bridging History to Current Operations

Springer

Editor
Peter L. Guth
Department of Oceanography
United States Naval Academy
Annapolis, MD, USA

ISSN 2522-8315 ISSN 2522-8323 (electronic)
Advances in Military Geosciences
ISBN 978-3-030-32172-7 ISBN 978-3-030-32173-4 (eBook)
https://doi.org/10.1007/978-3-030-32173-4

This Springer imprint is published by the registered company Springer Nature Switzerland AG
The registered company address is: Gewerbestrasse 11, 6330 Cham, Switzerland

Contents

Chapter 1
Introduction: Geosciences Supporting and Analyzing Military Operations

Peter L. Guth

The natural environment affects almost all human activity, including military operations. While scientific study of the earth sciences represents a relatively recent development, people have always noticed that the landscape, surface processes, climate, and weather affected what can or cannot be accomplished, including for warfare. They did not need soil science to understand that bogs would severely hamper mobility, or climatology to know that the range of possible activity changed between winter and summer, or the dry and wet seasons.

The International Association for Military Geosciences (IAMG) is a not-for-profit organization, administered by its members and represented by a Council. Membership in the Association is free and open to researchers and practitioners of military geology and geography and associated fields. Military geosciences, as defined by the IAMG, comprise all disciplines that are interested in military activities within a geological, geographical, or, more generally, spatial context. A few examples include engineering geology, hydrogeology, geospatial sciences, terrain analysis, archaeology, historical geography, geomorphology, environmental science, geophysics, climatology, and other cognate disciplines.

The IAMG has evolved in its format. Initial meetings included symposia at the annual meetings of geological societies in the USA and UK in 1994, 1996, 1998, and 2000. Starting in 2003, meetings solely hosted by IAMG have been held in 2003, 2005, 2007, 2009, 2011, 2013, 2015, 2017, and 2019 in the USA, Europe, and South Africa and have always featured field trips to observe the intersection of the landscape and the military. Many of the conferences have led to published books, including two in this series (Caldwell et al. 2004; McDonald and Bullard 2016). The papers included in this volume were initially presented at several of the conferences that did not produce a dedicated book.

P. L. Guth (✉)
Department of Oceanography, United States Naval Academy, Annapolis, MD, USA
e-mail: pguth@usna.edu

P. L. Guth (ed.), *Military Geoscience*, Advances in Military Geosciences,
https://doi.org/10.1007/978-3-030-32173-4_1

Papers presented at the conferences, and included in this volume, have a number of common themes:

- Studies of how the environment affects warfare, military training, military humanitarian operations, or environmental cleanup after military operations or training. Training at military academies still employs the campaigns of Alexander the Great, Hannibal, Napoleon, and the American Civil War to understand how military leaders in the past faced and overcame challenges. While these studies have a strong historical component, the focus on the environment makes them relevant today because our better technology still does not let us escape the impact of monsoons or dust storms. In fact modern technology like helicopters may be more affected by weather compared to foot operations.
- Studies on how the military used and employed geoscientists to support military training and operations. This could really only begin with the development of a professional scientific community in the nineteenth century, although prior to that time, miners and engineers had geotechnical skills that could be used to support military operations. The use of geoscientists changes over time and varies within different armed forces.
- Studies on how the environment will affect current and future military operations, from the tactical to the strategic levels. Modern technologies depend on raw materials like uranium, the rare earth elements, or lithium for batteries, and these are not evenly distributed and may be denied for many reasons. Climate change, sea level rise, and dams and river diversions will all impact military operations and bases in the coming decades. Politicians may try to ignore these changes, but business leaders, insurance companies, and military leaders know they must consider changing conditions and plan for them.
- Studies on how geographic tools like cartography, remote sensing, lidar data, and geographical information systems can assist military users. The military needs simple, easy to use tools that can assist soldiers in dangerous, fast-paced operations.

The chapters are arranged chronologically. The first three chapters cover the nineteenth century in Europe, the USA, and South Africa. Three chapters cover World War I and two chapters World War II. Four chapters cover late twentieth- and early twenty-first-century operations in the Middle East, while the last two chapters look toward the future and evolving needs.

References

Caldwell, D. R., Ehlen, J., & Harmon, R. S. (Eds.). (2004). *Studies in military geography and geology* (348 p). Dordrecht: Kluwer Academic Publishers. ISBN-13: 978-1402031045.

McDonald, E. V., & Bullard, T. (Eds.). (2016). *Military geosciences and desert warfare* (375 p). Springer. ISBN 978-1-4939-3429-4.

Chapter 2
Improvement in German Military Geology from the Napoleonic Wars to First World War

Hermann Häusler

Introduction

Military applications of geology first became apparent in Napoleonic times (Häusler 2003a, b; Häusler and Kohler 2003). When the French army entered Egypt in 1798, it was accompanied by a civilian Commission of Sciences and Arts that, among others, included Déodat de Dolomieu (1750–1801; Rose 2004, 2005a; Schramm 2006). These French mineralogists/geologists supported the army by exploring the geological resources of the country, rather than by contributing tactical or strategic advice (Rose 2005a), and therefore Betz (1984) mentioned that it "was not recognition of the military value of geology." Betz (1984) as well as Kiersch and Underwood (1998) credited not the French geologists involved in Napoleon's military campaign in Egypt but instead Karl von Raumer (1783–1865) with being an early military geologist when serving the Prussian General (later Field Marshal) Gebhard von Blücher in the Battle of Katzbach West of Breslau. Betz (1984) noted: "Probably the first time a geologist helped plan a military operation was in 1813, when the Prussian general von Blücher consulted Karl August von Raumer, a professor of mineralogy and one time student of Abraham Werner, for information on the terrain of Silesia in advance of the successful battle at Katzbach, in which Napoleon's army was defeated." Kiersch and Underwood (1998) agreed with him when they wrote: "The first documented military operation using geologic guidance was in 1813, when Professor K.A. von Raumer analyzed the terrain of Silesia for General von Blücher, which was significant to his defeat of Napoleon's Forces." However, nowhere in contemporary literature or in recognition of his merits as university professor, annotations on the tactical use of geologic information by von Blücher could be found. Karl Raumer, who personally did not take part in the Battle of Katzbach on August 26, 1813, gave more advice on Silesian mountains and rivers in general and served

H. Häusler (✉)
University of Vienna, Department of Environmental Geosciences, Vienna, Austria
e-mail: hermann.haeusler@univie.ac.at

P. L. Guth (ed.), *Military Geoscience*, Advances in Military Geosciences,
https://doi.org/10.1007/978-3-030-32173-4_2

as a riding courier in the German headquarters. However, in contrast to the ideas presented above, Häusler (2003b) and Rose (2005a, b) concluded that neither Captain Déodat de Dolomieu nor Captain Karl von Raumer initiated the military use of geology and that the pioneer use of military geology (in its stricter sense) should be credited to the Swiss-born Bavarian officer Johann Samuel von Gruner.

Johann Samuel von Gruner: Founder of Military Geology

Johann Samuel von Gruner was born in Bern, Switzerland, in 1766 and died near Munich, Germany, in 1824 (Table 2.1). Although he was not of noble birth, a special regulation of his hometown in 1783 allowed him to use the title "von." His name is known as (von) Gruner or Grouner, and both versions were used by him in published literature (Fig. 2.1). In this paper the spelling "Gruner" is preferred, though his fundamental paper on war geology is cited as Grouner (1826).

Gruner studied mining geology at the mining academy in Freiberg, Silesia, in 1787 (Fig. 2.2). From that time on, Gruner cultivated contacts with the famous

Table 2.1 Outline career of the Swiss geologist and officer Johann Samuel von Gruner, the forgotten father of German military geology

Feb 27, 1766	Born in Bern, Switzerland
1784–1785	Studies of geodesy and mining under Franz Samuel Wild in Bex at Swiss canton Waadt
1786–1791	Study of mining at Mining Academy Freiberg, Silesia under Abraham Gottlob Werner and excursions to France and Northern Italy
1794	Excursion to the Salzkammergut (Austria)
1796–1801	Co-editor of the atlas of Switzerland
1799	Participation in campaigns under General Lecourbe (as first-class captain = major)
Dec 6, 1802	Mining officer (Oberberghauptmann) of all Helvetic mines and salt mines
1803	Immigration to Bavaria
Dec 4, 1813	Gruner applies for Bavarian army
Jan 8, 1814	Promoted second-class captain
March 1814	Participation in campaigns against France
June 23, 1815	Promoted first-class captain (= major)
1815	Marriage to Clara Regina von Pallhausen
1820	Memorandum on the relationship between geology and war science ("Verhältnis der Geognosie zur Kriegswissenschaft"); not published
1821	Excursion to the Netherlands
Jan 31, 1824	Deadly road accident at Weilheim, Starnberg; Funeral in Munich.
1825	Posthumously published paper on the influence of geology on maps and reliefs ("Über den Einfluß der Geognosie auf Landcarten und Reliefs")
1826	Posthumously published paper on the relation of geology to war science ("Verhältnis der Geognosie zur Kriegs-Wissenschaft")

Fig. 2.1 Portrait of the Swiss Mining Officer Johann Samuel von Gruner dating 1800 (https://de.wikipedia.org/wiki/Johann_Samuel_von_Gruner) with signature "Joh: Sam: von Gruner, Oberberghauptmann"

academic teacher professor Abraham Gottlob Werner and two of his well-known students, the universal scientist and geographer Alexander von Humboldt and the geologist Leopold von Buch.

While serving in the French army as a major under General Claude-Jacques Lecourbe, in 1799 Gruner experienced the important influence of subsoil geology on military actions. At that time, Gruner was engaged in the preparation of a geographic atlas of Switzerland, which was then published in 1801. Furthermore, Gruner was involved in mining activities and finally became director general ("Directeur en Chef") of all Swiss mines. However, when Switzerland became the "Helvetic Republic" under Napoleon Bonaparte, Gruner lost his job. He immigrated to Bavaria in 1803.

At the beginning of his career in Munich, Gruner took care of the enhancement of optic instruments, which were necessary for the military geodetic survey of Bavaria. During the Liberation War, Gruner again served in the army, being raised to second-class captain in March 1814. The Bavarian king decorated Gruner for victoriously commanding a Bavarian infantry battalion during the advance to Paris against Napoleon I. During the last 10 years of his life, Gruner served as a

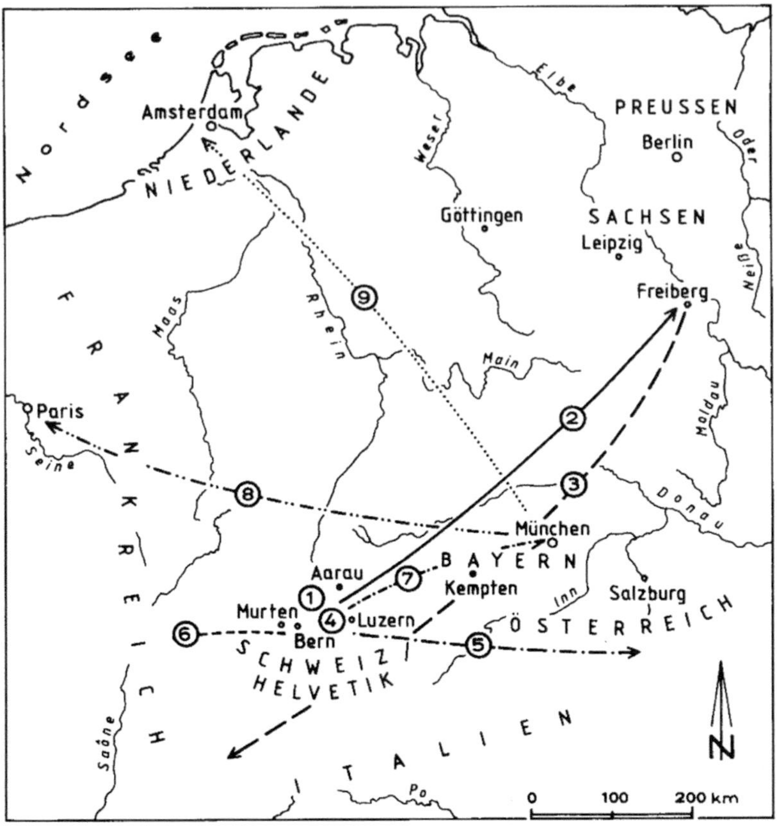

Fig. 2.2 Gruner's important journeys and stops in Europe between 1766 and 1821 (1–9, Häusler and Kohler 2003): (1) 1766–1785: Youth in Switzerland (2) 1787–1789: Mining Academy at Freiberg/Silesia (3) Excursions to France and Northern Italy (4) 1791–1803: Switzerland/Helvetic Republic (5) 1794: Excursion to Austria/Salzkammergut (6) 1799: Participation in the 2nd Coalition War (7) 1804: Immigration to Bavaria (8) 1814: Military campaign to Paris (9) 1821: Study trip to the Netherlands

commissioned officer and likely had close contact with General Clemens von Raglovich, chief of the Bavarian Bureau of Military Topography. Gruner wrote a memorandum on the relation of geology to war science ("Verhältnis der Geognosie zur Kriegs-Wissenschaft") in 1820, which was published posthumously in 1826 in "Moll's Neue Jahrbücher der Berg- und Hüttenkunde" (Grouner 1826; see Fig. 2.3 this paper). These mining yearbooks were published by Karl Maria Ehrenbert Freiherr von Moll who was a renowned statesman and scholar (Tichy 2003), and Gruner's paper became the fundamental publication on military geology in the early nineteenth century.

In the introduction of this paper, as shown in Fig. 2.3, Gruner states that "all sciences form a chain and not a one of them is superfluous. They are useful to well-informed persons of all professions, including educated officers." Among the other

Verhältnifs der Geognosie
zur
Kriegs-Wissenschaft
eine Skizze
von
*J. S. von Grouner *).*

Einleitung.

Alle Wissenschaften bilden eine Kette; sie bieten einander wechselseitig die Hand. Der gebildete Mann jeden Berufes kann sie alle trefflich benuzen; keine ist entberlich. Der gebildete Officier macht hievon keine Ausname.

Fig. 2.3 Title of the paper on war geology by Johann Samuel Gruner, which was published posthumously (Grouner 1826)

sciences, geology (at that time termed "Geognosie") was also important for the military – in particular for mountain and siege warfare. Gruner describes the advantages and disadvantages of strike and dip as well as of weathering of rock formations for mountain warfare in detail and refers to the disadvantages of higher rock permeability for the location of fortresses while also pointing out the counterbalancing advantages, such as springs. At this point it is concluded that Gruner was the very first geologist who was actively aware of the important influence of geology on military operations by the end of the seventeenth century.

When General Raglovich died, the knowledge on military geology diminished, and even General Carl von Clausewitz credited terrain for warfare in his famous book *On War* (Clausewitz 1832), however, ignored military geology. In 1854 the German geologist Bernhard von Cotta mentioned Gruner's paper on military geology in his book on the geology of Germany (Cotta 1854) before it was again neglected for decades. Further, Austro-Hungarian officer Baron Rudolf von Schmidburg (1810–1902) who published textbooks on terrain reconnaissance, first under his pseudonym "R.B.S" (Rudolf Baron Schmidburg – RBS 1855, 1869) and later under his full name (Schmidburg 1875, 1878, 1896), did not quote Gruner's paper and did not consider military geology at all in his numerous publications. Gruner's paper "Verhältnis der Geognosie zur Kriegs-Wissenschaft" was published in the last of the six mining yearbooks "Neue Jahrbücher der Berg- und Hüttenkunde" by the renowned statesman and natural scientist Karl Ehrenbert von Moll (1760–1838). Presumably this special yearbook was only used by mineralogists and geologists, and therefore it is small wonder that it remained unknown to the military.

Walter Kranz: Father of German Military Geology in Early Twentieth Century

Fifty years passed between Cotta quoting Gruner's pamphlet on military geology – nearly 100 years after the early military geologic experiences of the Swiss geologist and Bavarian officer Johann Samuel Gruner – before the German engineer officer Walter Kranz again published a paper on military geology in 1913. He was so convinced of the application of geology for fortification and its use for water supply and wastewater disposal that he published numerous papers on modern military geology both in geologic and military journals as well as in newspapers. Kranz alone put his stamp on military geology from the First World War and until the beginning of the Second World War; he can justifiably be termed the father of military geology of the early twentieth century (Fig. 2.4).

Walter Kranz already was commissioned as an engineer officer when he studied geology at universities of Strasbourg, Bonn, and Greifswald from 1901 to 1909. During that time he published his first geologic paper and was particularly interested in the formation of the crater of the "Nördlinger Ries" about which he published some 39 papers (e.g., Kranz 1911). In 1913 Kranz was appointed to the Fortification Service of the Strasbourg Fortress when he published his first article on

Fig. 2.4 The engineer officer and geologist Dr. Walter Kranz, father of German military geology in the early twentieth century (Häusler 2003b)

military geology in "Kriegstechnische Zeitschrift," the German journal on war technology (Kranz 1913). Comparable to Gruner's introduction of his paper on war geology (Grouner 1826), Kranz (1913) also noted in the introduction of his paper on military geology that "the sciences influence all spheres of military affairs, and some of them have taken permanent residence in military administrative offices, as has e.g. electro-physics in Department III of the Royal Prussian Engineer-Committee. Only one natural science is today still regarded as a stepchild in most military matters: geology. The degree of a weapon's effectiveness being dependent on careful adaptation to the terrain in any form of warfare, the soldier becomes that much more sensitive to the influence of the ground itself and can less and less ignore the teachings of the underground/subsurface – geology" (Fig. 2.5). This was explained in detail using examples from trench warfare, fortification, fortification warfare, mining warfare, water supply of troops, and construction of military roads and railways. As a consequence, Kranz emphasized basic courses for officers in petrography, tectonics, geohazards such as earthquakes and mass movements, and applied geology including geologic mapping. Kranz proposed that persons with experience similar to his own train engineer officers in military geology.

It cannot be ruled out that Walter Kranz was aware of Gruner's paper on military geology published in 1826 prior to writing his 1913 article because in his contribution to the development of war geology (Kranz 1920b), Kranz referred to Gruner's military geology experiences during the Napoleonic Wars. In 1914, again comparable to Gruner's 1820 memorandum on the relationship between geology and war science (Table 2.1), Captain Kranz wrote his own memorandum on the need for military geologists for fortification and preparation of fortification war (Table 2.2). In August 1914 Kranz became the geological advisor of the 6th Army High Command, and at that time the only other military geologist was the geographer Dr. Siegfried Passarge, who was deployed as a military geologist to the 4th Army High Command at the western front. As a consequence of Kranz's memorandum in April 1915, the Prussian engineer generals of the German army high commands were advised to make use of geologists, and by November 1915 six geologic units of the

Militärgeologie.

Von W. Kranz, Hauptmann z. D., zugeteilt der Fortifikation Straßburg i. E.

I. Aufgaben.

In fast alle Gebiete des Militärwesens dringen allmählich die Wissenschaften vor, und manche davon haben sich in militärischen Dienststellen dauernden Sitz erobert, wie z. B. die Elektrophysik in Abteilung III des Königlich preußischen Ingenieur-Komitees. Nur ein Gebiet der Naturwissenschaften muß heute noch in den meisten militärischen Dingen als Stiefkind beiseite stehen: Die Geologie.

Je mehr die Waffenwirkung zur sorgfältigen Anpassung an das Gelände in jeder Kriegsform zwingt, um so fühlbarer wird dem Soldaten der Einfluß des Bodens selbst, um so weniger darf er der Lehre vom Erdboden, der Geologie, gleichgültig gegenüberstehen. Der neuerdings stetig an Be-

Fig. 2.5 Introduction to the fundamental paper on military geology published by Captain Walter Kranz in the Journal on War Technique ("Kriegstechnische Zeitschrift") in 1913 (Kranz 1913)

Table 2.2 Outline career of the officer and geologist Dr. Walter Kranz, pioneer of German military geology (Kranz 1917a; Rose et al. 2000; Häusler 2003b)

18 Apr. 1873	Born in Wesel, Niederrhein
27 Jan. 1893	2nd Lieutenant in the Engineer Battalion15, at Strasbourg
1895–1896	"Vereinigte Artillerie- und Ingenieurschule", Berlin
27 Jan.1902	First lieutenant (Oberleutnant)
1902–1903	Engineer Battalion 13, at Ulm
1901–1909	Study of geology at universities of Strasbourg, Bonn and Greifswald
1903–1905	Engineer Inspectorate, with service in Neubreisach Fortress
16 Oct.1906	Captain; in Lotharingian Engineer Battalion 20, at Metz
1909–1912	Engineer Inspectorate, with service at Swinemünde
22 Mar. 1913	Captain z. D. (zum Dienst = on duty); appointed to the Fortification Service of the Strasbourg Fortress
1913	Paper on military geology published in "Kriegstechnische Zeitschrift"
10 Mar. 1914	Memorandum on the need of military geologists for fortification and preparation of fortification war
1 Aug. 1914	Geological advisor of the 6th Army High Command (Armeeoberkommando 6 = AOK6)
28 Mar. 1916	Promoted Major
14 Nov. 1916	Submission of PhD at Munich University "Über Boden-Filtration, Lage und Schutz von Wasserfassungen, mit besonderer Berücksichtigung militärischer Erfordernisse"
Jan. 1917	Geologic department of Strasbourg Fortification
Mar. – June 1917	War geologist in the geology group of the Geodetic Department 13 (Württemberg) of the Armee-Abteilung B
1917	Marriage to Hanna Heineken († 1926)
1919	Discharged in Strasbourg as reserve army major
16 June 1919	Employed as a state geologist at the Geologic Department of the Statistic State Office of Baden-Württemberg
	University lecturer on palaeontology, civil engineering and military geology at Stuttgart University
1929	Marriage to Anneliese Rode († 1949)
1938	Retired as state geologist
30 Dec. 1953	Died at Bopfingen, west of Nördlingen

army headquarters (commanded by General von Strantz) at the western front comprised 28 war geologists. However, it took until July 1916 for war geologists to be officially deployed to the geodetic departments ("Vermessungsabteilungen") of the Royal Prussian Topographic Survey (Häusler 2000).

In March 1916 Kranz was promoted to major, and in November 1916 he submitted a paper on "Soil filtration, position and protection of water catchments with particular consideration of military requirements" and received his PhD in geology

Über Boden-Filtration,
Lage und Schutz von Wasserfassungen,
mit besonderer Berücksichtigung militärischer Erfordernisse.

Von

Major z. D. W. Kranz.

Mit 7 Textfiguren.

Inaugural-Dissertation

der

Philosophischen Fakultät (Sektion II) der Ludwig-Maximilians-
Universität München,

eingereicht am 14. 11. 1916.

STUTTGART 1917.
E. Schweizerbart'sche Verlagsbuchhandlung
Nägele & Dr. Sproesser.

Fig. 2.6 Title of the PhD thesis of Major Walter Kranz on “Soil filtration, position and protection of water catchments with particular consideration of military requirements” submitted to the University of Munich in November 1916, while Kranz was commissioned engineer officer at the Strasbourg Fortress (Kranz 1917a)

at the Philosophical Faculty of the Ludwig Maximilian University of Munich (Kranz 1917a, b, c, d; see Fig. 2.6 this paper). Later on, Kranz served as a war geologist in France and Flanders; from March to June 1917, he was deployed to the geology group of the Geodetic Department 13 (Württemberg) of “Armee-Abteilung B” (Häusler 2000). Discharged in Strasbourg as a reserve major in 1919, Kranz was employed as a state geologist in Württemberg and also as a university lecturer in paleontology, construction engineering, and military geology (Rose et al. 2000; Table 2.2 this paper).

Major Kranz set a milestone in military geology, matching applied geology with military needs. However, despite his early success, the German army high command was not aware of the use of military geologists at the beginning of the First World War. During four long and awful years of fighting, and despite previous strategic assumptions, rapid battles were followed by endless sapping and military mining activities. Nonetheless, the German army did not establish an effective military geology organization until the end of the First World War.

Kranz was the first military geologist of the early twentieth century who also informed the public about the use of military geologists in times of both war and peace when he published articles in daily newspapers such as the "Straßburger Post" or in the "Schwäbischer Merkur." In total, Kranz wrote about 170 publications, approximately one third of which dealt with his experiences in military geology.

During the First World War, Kranz published applied geological and military geological papers on the following subjects:

- General overview on the new responsibilities of military geologists in peace and war (Kranz 1914, 1915a, b, c, d, e)
- Geology and hygienic conditions in trench warfare (Kranz 1916a, b, c)
- Drinking water supply and water treatment (Kranz 1917a, b, c)
- Provision of raw materials for military use (Kranz 1917d)

After the First World War, Major Kranz published his experiences as a war geologist in papers on the following subjects:

- Development of war geology in the First World War (Kranz 1920b, 1927a, b, 1933b, 1934a, b, c, 1935a, b, d, 1936d)
- Provision of raw materials (Kranz 1919b)
- Special geologic investigation methods (Kranz 1920a)
- Blasting techniques (Kranz 1928, 1929, 1931a, 1933a)
- Mining in the First World War (Kranz 1936a, b, c, 1936e, 1937a)
- Water supply in general (Kranz 1918, 1919a)
- Water supply of troops (Kranz 1919b, 1936f, 1937b; Kranz and Scupin 1937)
- Use of the divining rod for water supply (Kranz 1921, 1922a, b, 1930, 1931b, c, 1932)
- Review of foreign military geology papers
- Military geology, termed "Wehrgeologie" (Kranz 1935c, 1936g, 1937c, d, 1938a, b, c, d, 1939a, b, c)

Like Walter Kranz, Ernst Wochinger also completed his PhD in military geology in 1917 while fighting in the First World War. Ernst Wochinger served as an engineer officer and also recognized the importance of geology for military use. Interestingly, both Wochinger and Kranz studied applied geology during the war in Munich – Wochinger at the Royal Technische Hochschule and Kranz at the Ludwig-Maximilians-Universität (Fig. 2.7).

Beitrag zur Geschichte der Ingenieurgeologie

unter

besonderer Berücksichtigung der Kriegsgeologie.

Von der K. technischen Hochschule zu München zur Erlangung der Würde eines Doktors der technischen Wissenschaften (Doktor-Ingenieurs) genehmigte Dissertation.

Vorgelegt von

Diplom-Ingenieur Ernst Wochinger aus Regensburg.

Referent:
Geheimer Hofrat Professor
Dr. Konrad Oebbeke.

Korreferent:
Professor Dr. Konrad Pressel.

Angenommen: München 17. Dezember 1917.

Traunstein 1919.
Buchdruckerei Ed. Leopoldseder, Traunstein.

Fig. 2.7 First page of the published dissertational thesis of Ernst Wochinger on the "Contribution to the history of engineering geology with special reference to war geology" (Wochinger 1919)

Wochinger's dissertation comprised chapters on the history of war geology, examples of war geology, and comments on organization and education. Wochinger's military geology activities were supported at the Royal Technical University in Munich by Univ. Prof. Dr. Konrad Oebbeke and by Dr. Leopold van Werveke. Werveke was the director of the Geological Survey of Elsaß-Lothringen; he wrote a short contribution on war geology in a newspaper. In March 1918 Oebbeke was deployed as a war geologist at the western front. In contrast to Kranz, Ernst Wochinger neither served the German army as a war geologist nor did he publish

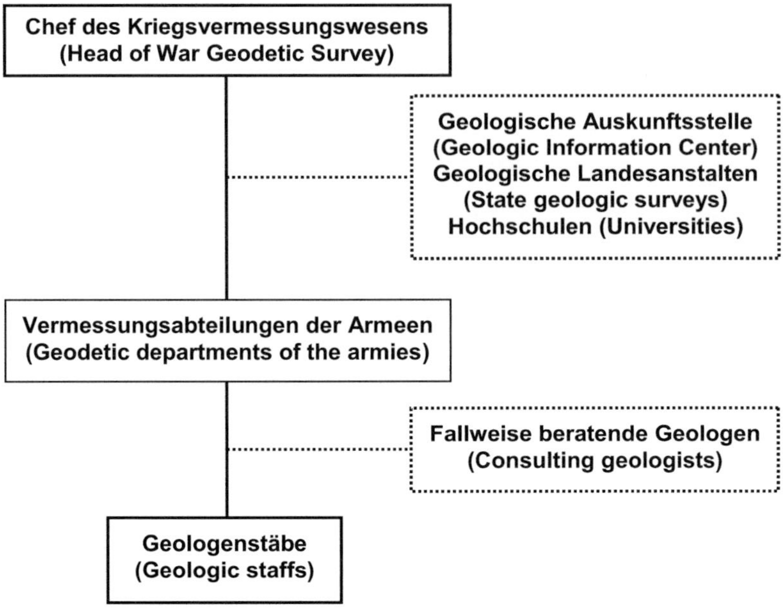

Fig. 2.8 Organigram of German military geology in 1917 which was temporarily supported by civil institutions and consulting geologists (Wochinger 1919)

specific papers on military geology despite summarizing the organization of war geology in his thesis. From July 4, 1915, until the end of the war, major of the general staff Siegfried Boelke headed the War Geodetic Office which comprised the army's geodetic departments which in turn consisted of geologic staffs. Occasionally the military organization of war geology was supported by civil institutions such as provincial geologic surveys, universities, or consulting geologists (Wochinger 1919, see Fig. 2.8 this paper).

The history of the First World War reveals that the engagement of Major Kranz in military geology was successful, despite the slow implementation. Although there were only two military geologists in the German army in August 1914, by the end of 1918, a geology unit was attached to each of the German army high commands so that in total around 200 war geologists served the 28 geodetic departments of the army. About 50 different leaflets were printed for information, 13 different war geologic maps were provided at 1:10,000 scale and 11 more at 1:25,000 scale, and in total about 5500 geological expertises supported the military actions of the German armies (Häusler 2000). Despite retiring as state geologist in 1938, Kranz still remained engaged in military geology. In 1938 he published a very popular textbook on "technical military geology" which aimed to be a companion for soldiers, geologists, technicians, medical doctors, and other experts (Kranz 1938a). In addition, Kranz was co-author of a textbook on "military geology" (Bülow et al. 1938), and later on he published another paper on technical military geology in the *Journal of Applied Geology* (Kranz 1943, Häusler 1995a, b).

Acknowledgments The author thanks two anonymous reviewers for constructive and thorough reviews.

References

Betz, F. (1984). Military geoscience. In C. W. Finkl Jr. (Ed.), *The encyclopedia of applied geology* (Encyclopedia of earth sciences) (Vol. XIII, pp. 355–359). New York: van Nostrand Reinhold.

Bülow, K., Kranz, W., & Sonne, E. (1938). *Wehrgeologie* (p. 170). Leipzig: Quelle & Meyer, Leipzig.

Clausewitz, M. (1832). *Hinterlassene Werke des Generals Carl von Clausewitz über Krieg und Kriegsführung 1. Band: Vom Kriege (On war)* (p. 371). Berlin: Ferdinand Dümmler.

Cotta, B. (1854). *Deutschlands Boden und sein geologischer Aufbau und dessen Einwirkungen auf das Leben der Menschen (The soil of Germany, its geological composition and effects on the life of humans). Erste Abtheilung* (p. 614). Leipzig: Brockhaus.

Grouner, J. S. (1826). Verhältnis der Geognosie zur Kriegs-Wissenschaft (The relation of geology to war science). *Moll's Neue Jahrbücher der Berg- und Hüttenkunde, 6*(2), 187–233.

Häusler, H. (1995a). *Die Wehrgeologie im Rahmen der Deutschen Wehrmacht und Kriegswirtschaft. Teil 1: Entwicklung und Organisation (Military geology of the German Wehrmacht and the economy of war. Part 1: Development and organisation). Informationen des Militärischen Geo-Dienstes 47* (p. 155). Wien: Bundesministerium für Landesverteidigung.

Häusler, H. (1995b). *Die Wehrgeologie im Rahmen der Deutschen Wehrmacht und Kriegswirtschaft. Teil 2: Verzeichnis der Wehrgeologen (Military geology of the German Wehrmacht and the economy of war. Part 2: Register of military geologists). Informationen des Militärischen Geo-Dienstes 48* (p. 119). Wien: Bundesministerium für Landesverteidigung.

Häusler, H. (2000). *Die Österreichische und Deutsche Kriegsgeologie 1914–1918 (Austrian and German war geology 1914–1918). Informationen des Militärischen Geo-Dienstes 75* (p. 161). Wien: Bundesministerium für Landesverteidigung.

Häusler, H. (2003a). *Johann Samuel Gruner (1766–1824) – Begründer der Militärgeologie (Johann Samuel Gruner – Founder of military geology). MILGEO 9* (p. 50). Wien: Bundesministerium für Landesverteidigung.

Häusler, H. (2003b). *Dr. Walter Kranz (1873–1953) – der erste Militärgeologe des 20. Jahrhunderts (Dr. Walter Kranz – The first military geologist of the 20th century). MILGEO 12* (p. 80). Wien: Bundesministerium für Landesverteidigung.

Häusler, H., & Kohler, E. E. (2003). Der Schweizer Geologe, Oberberghauptmann und Major Johann Samuel Gruner (1766–1824) – Begründer der Militärgeologie (The Swiss geologist, mining officer and Major Johann Samuel Gruner – Founder of military geology). Minaria Helvetica 23a: 47–102.

Kiersch, G. A., & Underwood, J. R., Jr. (1998). Geology and military operations, 1800–1960: An overview. In: Underwood JR Jr, Guth PL (eds) Military geology in war and peace. *Reviews in Engineering Geology, 13*, 5–27.

Kranz, W. (1911). Das Nördlinger Riesproblem I (The problem of the Nördlinger Ries crater I). *Jahresberichte und Mitteilungen des oberrheinischen geologischen Vereins NF, 1*, 32–35.

Kranz, W. (1913). Militärgeologie (Military geology). *Kriegstechnische Zeitschrift, 16*, 464–471.

Kranz, W. (1914). Militär-Geologie (Military geology). Straßburger Post Nr. 163, Mittagsausgabe.

Kranz, W. (1915a). Kriegsgeologie (War geology). Straßburger Post Nr. 173, Mittagsausgabe.

Kranz, W. (1915b). Aufgaben der Geologie im Mitteleuropäischen Kriege (The relevance of geology in the Middle European War). *Petermanns Mitteilungen, 61*, 249–255.

Kranz, W. (1915c). Aufgaben der Geologie an der Kampffront im mitteleuropäischen Kriege (The relevance of geology on the front line in the Middle European War). *Mitteilungen über Gegenstände des Artillerie- und Geniewesens, 46*, 989–996.

Kranz, W. (1915d). Aufgaben der Geologie an der Kampffront im mitteleuropäischen Kriege (The relevance of geology on the front line during the Middle European War). *Mitteilungen über Gegenstände des Artillerie- und Geniewesens, 46*, 1120–1127.

Kranz, W. (1915e). Kriegsgeologie (War geology). *Der Geologe, 14*, 1–2.

Kranz, W. (1916a). Geologie und Hygiene im Stellungskrieg. I. Die Kriegs-Sanitätsordnung; II. Der Abessinierbrunnen (Geology and hygiene in trench warfare. I. The war sanitary policy; II The Abessinier well). Centralblatt für Mineralogie. *Geologie und Paläontologie, 1916*, 270–276.

Kranz, W. (1916b). Geologie und Hygiene im Stellungskrieg. III. Brunnenbohrungen (Geology and hygiene in trench warfare; III. Well drilling). Centralblatt für Mineralogie. *Geologie und Paläontologie, 1916*, 291–356.

Kranz, W. (1916c). *Geologie und Hygiene im Stellungskrieg (Geology and hygiene in trench warfare)* (p. 73). Stuttgart: Schweizerbarth.

Kranz, W. (1917a). *Über Boden-Filtrationen, Lage und Schutz von Wasserfassungen, mit besonderer Berücksichtigung militärischer Erfordernisse (On soil-filtration, position and protection of water catchments with special reference to military use). Inaugural-Dissertation der Philosophischen Fakultät (Sektion II) der Ludwig-Maximilians-Universität München* (p. 62). Stuttgart: E. Schweizerbart.

Kranz, W. (1917b). Wasserversorgung durch offene Gräben, Sickerung, Drainage (Water supply by open trenches, infiltration and drainage). *Zeitschrift für praktische Geologie, 25*, 1–22.

Kranz, W. (1917c). Künstliche Trinkwasserbereitung und Verbesserung (Artificial preparation and purification of drinking water). *Internationale Zeitschrift für Wasserversorgung, 4*, 12–15.

Kranz, W. (1917d). Beschaffung von Rohstoffen des Bodens für militärische Erfordernisse (Provision of raw materials for military use). *Zeitschrift für praktische Geologie, 25*, 59–66.

Kranz, W. (1918). Grundwasser und Quellen (Groundwater and springs). *Neues Jahrbuch für Mineralogie, Geologie und Paläontologie, 1918*, 43–48.

Kranz, W. (1919a). Zur Sozialisierung der Wasserversorgung des Grundwassers und der Quellen (Socialisation of drinking water supply from groundwater and wells). *Naturwissenschaftliche Wochenschrift NF, 18*, 312–317.

Kranz, W. (1919b). Die Geologie in der Kriegs-Literatur, bei Beschaffung von Rohstoffen des Bodens und Wasserversorgung für Truppen (Geology in war literature, for provision of raw materials and for water supply of troops). In B. Schmid (Ed.), *Deutsche Naturwissenschaft* (pp. 427–454). München: Technik und Erfindung im Weltkriege. O. Nemnich.

Kranz, W. (1920a). Ergebnisse geologischer Untersuchungsmethoden bei Beton-Druckproben (Results of geological testing methods for use in concrete pressure testing). *Zeitschrift für praktische Geologie, 28*, 101–127.

Kranz, W. (1920b). Beiträge zur Entwicklung der Kriegsgeologie (Teil I: Einleitung; Teil II: England) (Contributions to the development of war geology. Part I: Introduction; Part II: England). *Geologische Rundschau, 11*, 329–349.

Kranz, W. (1921). Beiträge zur Klärung der Wünschelrutenfrage (Contributions to solving the question of the divining rod). *Zeitschrift für praktische Geologie, 29*, 171.

Kranz, W. (1922a). Beiträge zur Klärung der Wünschelrutenfrage (Contributions to solving the question of the divining rod). *Zeitschrift für praktische Geologie, 30*, 41–47.

Kranz, W. (1922b). Beiträge zur Klärung der Wünschelrutenfrage (Contributions to solving the question of the divining rod). *Zeitschrift für praktische Geologie, 30*, 53–58.

Kranz, W. (1927a). *Die Geologie im Ingenieur-Baufach (Geology for civil engineering)* (p. 425). Stuttgart: Enke.

Kranz, W. (1927b). Die Entwicklung der Kriegsgeologie und ihre Bedeutung für die allgemeine Angewandte Geologie (Development of war geology and its relevance to applied geology in general). In W. Kranz (Ed.), *Die Geologie im Ingenieur-Baufach (Geology for civil engineering)* (p. 52). Stuttgart: Enke.

Kranz, W. (1928). Sprengtheorie (Theory of blasting). *Zeitschrift der Deutschen Geologischen Gesellschaft, 80*, 1–257.

Kranz, W. (1929). Vulkanexplosionen, Sprengtechnik, praktische Geologie und Ballistik (Vulcanic explosions, blasting technique, applied geology and ballistics). *Zeitschrift der Deutschen Geologischen Gesellschaft, 80*, 257–307.

Kranz, W. (1930). Zur Tektonik des Blattgebietes Weissach, mit Beitrag zur Wünschelrutenfrage (Tectonics of the geological map Weissach, with a contribution to the question of the divining rod). *Zeitschrift der Deutschen Geologischen Gesellschaft, 82*, 117–124.

Kranz, W. (1931a). Bodenerschütterungen, Geophysik und Ingenieurgeologie (Ground vibrations, geophysics and engineering geology). *Zeitschrift für praktische Geologie, 39*, 38–40.

Kranz, W. (1931b). Ein wirklicher geologischer, hydrologischer und Wünschelrutenerfolg beim Mangfall-Stollenbau (A real geologic, hydrologic and divining rod success during the construction of the Mangfall Tunnel). *Centralblatt für Mineralogie, Geologie und Paläontologie, Abt.B*, 376–382.

Kranz, W. (1931c). Wünschelrutenversuche am Gebirgsbau der Blattgebiete Leonberg und Weissach NW Stuttgart, mit Beitrag zur Baugrundgeologie (Experiments with the divining rod in the area of the geological maps northwest of Stuttgart, with a contribution to foundation geology). *Archiv zur Klärung der Wünschelrutenfrage, 1*, 3–42.

Kranz, W. (1932). Bohrungen und Wünschelrute (Drillings and the divining rod). *Zeitschrift für praktische Geologie, 40*, 125–127.

Kranz, W. (1933a). Die Sprengtechnik als Werkzeug der geologischen Forschung (Blasting technique as a tool for geologic investigation). *Zeitschrift des Vereins Deutscher Ingenieure, 77*, 850.

Kranz, W. (1933b). Kriegsgeologische Erkundungen für Angriff und Verteidigung (War geologic reconnaissance for attack and defence). *Taktik und Technik, Beilage zu Deutsche Wehr, 37*(16), 61–63.

Kranz, W. (1934a). Beiträge zur Entwicklung der Kriegsgeologie (Fortsetzung, III. Rußland) (Contribution to the development of war geology: Continuation, III. Russia). *Geologische Rundschau, 25*, 194–201.

Kranz, W. (1934b). Kriegsgeologische Erkundungen für Angriff und Verteidigung (War geologic reconnaissance for attack and defence). *Deutsche Pionier-Zeitung*, (3–6), 18, 26–27, 34, 44.

Kranz, W. (1934c). Russische Kriegs- und Militärgeologie (Russian war- and military geology). *Schwäbischer Merkur, 1934*(108), Sonntagsbeilage 13.5.1934.

Kranz, W. (1935a). Minierkampf und Kriegsgeologie an der Westfront 1914–18 (Mining conflicts and war geology at the Western Front 1914–18). *Sonderbeilage zu Deutsche Wehr, 39*(47), 2–8.

Kranz, W. (1935b). Minierkampf und Kriegsgeologie im Wytschaetebogen (Mining conflicts and war geology at the Wytschaete Ridge). *Vierteljahreshefte für Pioniere, 2*, 166–181.

Kranz, W. (1935c). Wehrgeologie und Pionier-Sprengdienst (Military geology and blasting-engineer service). *Wehrtechnische Monatshefte, 39*, 2–10.

Kranz, W. (1935d). Der Geologe bei der Feldbefestigung (Geology for field fortification). *Deutsche Pionier-Zeitung, 1*, 5.

Kranz, W. (1936a). Allgemeine Lehren aus den Minierkämpfen im Weltkrieg: Die Wytschaete-Sprengungen bei Hollandscheschuur und Petit Bois; taktische Wirkungen der Sprengungen am 7. Juni 1917 (General conclusions from mining conflicts during the Great War: The Wytschaete-blastings at Hollandscheschuur and Petit Bois; tactical outcomes of the explosions on June 7, 1917). *Sonderbeilage zu Deutsche Wehr, 40*(31), 1–6.

Kranz, W. (1936b). Feldminierkampf und Kriegsgeologie. Monte Cimone – Col di Lana – Monte Sief (Field mining conflicts and war geology. Monte Cimone – Col di Lana – Monte Sief). *Zeitschrift für das gesamte Schieß- und Sprengstoffwesen, 31*(3,4), 73–75, 118–121.

Kranz, W. (1936c). Kämpfe im Wytschaetebogen und Kriegsgeologie (Battles at Wytschaete Ridge and war geology). *Nachrichten aus Württemberg, 1936*, 209.

Kranz, W. (1936d). Militärgeologische und technische Organisationsfragen (Questions of military geologic and technical organisation). *Wehrtechnische Monatshefte, 40*, 312–318.

Kranz, W. (1936e). Minierkampf und Kriegsgeologie am Monte Pasubio 1916–18 (Mining engagements and war geology at Monte Pasubio 1916–18). *Wehrtechnische Monatshefte, 39*, 7–13.

Kranz, W. (1936f). Truppenwasserversorgung, Kriegsgeologie -Technik und Hygiene (Troop water supply, war geology – Technique and hygiene). *Deutsche Wehr, 40*, 91–92.

Kranz, W. (1936g). Wehrgeologie, Pioniersprengdienst und Minierkampf (Military geology, blasting-engineer service and mining conflicts). *Zeitschrift für Berg-, Hütten- und Salinenwesen im Deutschen Reich, 84*, 65–67.

Kranz, W. (1937a). Der Kampf um Wytschaete (The battle of Wytschaete). *Deutsche Pionier-Zeitung, 14*, 13.

Kranz, W. (1937b). Kriegsgeologie und Truppenwasserversorgung an der Belgischen Küste 1915 (War geology and troop water supply along the Belgian coast 1915). *Vierteljahreshefte für Pioniere, 4*(2), 96–99.

Kranz, W. (1937c). Wehrgeologie in ihrer Bedeutung für die Landesverteidigung (Military geology and its implication for national defence). *Wehrtechnische Monatshefte, 41*, 327–331.

Kranz, W. (1937d). Wehrgeologie und ihre Bedeutung für das Heer (Military geology and its implications for the army). In H. Franke (Ed.), *Handbuch der neuzeitlichen Wehrwissenschaften, II. Bd. "Das Heer" (Handbook of modern defence sciences, volume II: "The army")* (pp. 782–788). Berlin: de Gruyter.

Kranz, W. (1938a). *Technische Wehrgeologie. Wegweiser für Soldaten, Geologen, Techniker, Ärzte, Chemiker und andere Fachleute (Technical military geology. Guide for soldiers, geologists, technicians, medical doctors and other experts)* (p. 78). Leipzig: Jänecke.

Kranz, W. (1938b). Rezension von: Bülow K, Kranz W, Sonne E, Burre O, Dienemann W (1938) Wehrgeologie (Review of: Bülow K, Kranz W, Sonne E, Burre O, Dienemann W (1938) Military geology). *Geologische Rundschau, 29*, 99–101.

Kranz, W. (1938c). Was ist "Wehrgeologie"? (What is military geology?). *Zeitschrift für praktische Geologie, 46*, 35.

Kranz, W. (1938d). Wehrgeologie (Military geology). In: H. Franke (Ed.), *Handbuch der neuzeitlichen Wehrwissenschaften, III. Bd. "Kriegsmarine" und "Luftwaffe" (Handbook of modern defence sciences, volume III: "War Marine" and "Airforce")* (pp. 429–431, pp. 438–440). Berlin: de Gruyter.

Kranz, W. (1939a). Straßen- und Wasserbaupioniere u. Wehrgeologen (Road construction and hydraulic engineers and military geologists). *Deutsche Pionier- Zeitung, 17*, 181–182.

Kranz, W. (1939b). Wehrgeologisch-technische Baustoffbeschaffung im Kampfgebiet (Military geological and -technical construction material provision in the battle field). *Wehrtechnische Monatshefte, 43*, 399–407.

Kranz, W. (1939c). Wehrgeologisch-technische Kartierung (Military geologic technical mapping). *Deutsche Wehr, 43*(14), 257–258.

Kranz, W. (1943). Zur Entwicklung der Deutschen Technischen Wehrgeologie (On the development of German technical military geology). *Zeitschrift für praktische Geologie, 51*, 91–92.

Kranz, W., & Scupin, H. (1937). Minierschutz und Trinkwasserversorgung im Karpatenkorps 1917 (Mining protection and drinking water supply in the Carpathian Corps 1917). *Vierteljahreshefte für Pioniere, 4*, 240.

RBS. (1855). *Grundzüge einer physikalisch vergleichenden Terrainlehre in ihrer Beziehung auf das Kriegswesen als Leitfaden zum Vortrage und Selbstunterrichte für Eingeweihte und Laien (Fundamentals of physically comparable terrain reconnaissance with regard to warfare as a guide for lectures and self-instruction for the experienced as well as laymen)* (p. 324). Agram: National-Buchdruckerei.

RBS. (1869). *Grundzüge einer physikalisch vergleichenden Terrainlehre in ihrer Beziehung auf das Kriegswesen (Fundamentals of physically comparable terrain reconnaissance with regard to warfare)* (2. verbesserte und vermehrte ed., p. 279). Wien: Gerold's Sohn.

Rose, E. P. F. (2004). Napoleon Bonaparte's Egyptian campaign of 1798: The first military operation assisted by geologists? *Geology Today, 20*(1), 24–29.

Rose, E. P. F. (2005a). Military geology. In R. C. Selley, L. R. M. Cocks, & I. R. Plimer (Eds.), *Encyclopedia of geology* (Vol. 3, pp. 475–487). Oxford: Elsevier.

Rose, E. P. F. (2005b). Karl von Raumer: A pioneer of German military geology. *Geology Today, 21*(5), 182–186.

Rose, E. P. F., Häusler, H., & Willig, D. (2000). Comparison of British and German applications of geology in world war. In R. EPF & C. P. Nathanail (Eds.), *Geology and warfare: Examples of the influence of terrain and geologists on military operations* (pp. 107–140). London: The Geological Society of London.

Schramm, J. M. (2006). *Geologie & Untergrund – das Operationsfeld der Militärgeologie. (Geology and the underground – The area of operations for military geology) MILGEO 8* (p. 208). Wien: Bundesministerium für Landesverteidigung.

Tichy, G. (2003). Karl Maria Ehrenbert Freiherr von Moll. Staatsmann und Gelehrter (1760–1838). (Karl Maria Ehrenbert Freiherr von Moll. Statesman and scholar). *Berichte der Geologischen Bundesanstalt, 64*, 82–84.

von Schmidburg, R. (1875). *Grundzüge einer physikalisch-vergleichenden Terrainlehre in ihrer Beziehung auf das Kriegswesen. 3. auf das metrische Maß umgearbeitete und wesentlich bereicherte Auflage. Fundamentals of physically comparable terrain reconnaissance with regard to warfare* (3. Transposed to the metric system and significantly expanded ed., p. 378). Wien: Gerold's Sohn.

von Schmidburg, R. (1878). *Physikalisch-vergleichende Terrainlehre in ihrer Beziehung auf das Kriegswesen (Physically comparable terrain reconnaissance with regard to warfare)* (4. vervollst. ed., p. 440). Wien: Gerold's Sohn.

von Schmidburg, R. (1896). Anleitung zur Orientierung im Gebirge nach den Grundlinien der in die physikalisch vergleichende Terrainlehre eingreifenden Geologie der Gegenwart. Als Behelf zur schnelleren Auffassung des Gebirgszusammenhanges bei Ausführung militärischer Terrain-Recognoscirungen (Instruction for orientation in mountainous regions based on the principles of present day geology as it treats physically comparable terrain reconnaissance. Assistance for faster comprehension of the relationship between mountain-formations by using military terrain reconnaissance). Leykam, Graz, p. 446.

Wochinger, E. (1919). Beitrag zur Geschichte der Ingenieurgeologie unter besonderer Berücksichtigung der Kriegsgeologie (Contribution to the history of engineering geology with special consideration of war geology). Veröffentlichte Dissertation der königlich technischen Hochschule München. Leopoldseder, Traunstein, p. 164.

Chapter 3
The Battle of Big Hole (Montana, USA, 1877): History, Archaeology, GIS, and Military Terrain Analysis

Peter L. Guth and Douglas D. Scott

Introduction

Understanding the past requires the integration of a number of tools and approaches. The analysis of historical battles and battlefields provides one of the best examples of the utility of this approach, with important contributions from historical records and documents, and remains recovered by battlefield archaeology. The application of military terrain analysis to understand how the commanders used the physical environment and how it affected the outcomes adds depth to this understanding. Geographical information systems (GIS) can integrate the collected data and the geography of the battlefield to help in interpretation, analysis, and display of the results. High-quality, free digital map data allows us to combine the archaeological data with a military terrain analysis to better understand the battle.

The Battle of Big Hole, which was part of the Nez Perce War in the summer of 1877, took place in southwest Montana, not far from the Idaho border (Fig. 3.1). It provides an outstanding case study of how a GIS analysis can contribute to our understanding of a historic battle. A wealth of historic documents and eyewitness accounts provide a record from both sides (summarized in Haines 1999; Greene 2000). A comprehensive archaeological survey and forensic analysis (Scott 1994) document many of the troop movements, using the methodology first perfected at the Custer Battlefield at Little Bighorn, Montana, in June 1876 (Fox 2003; Scott et al. 2000).

P. L. Guth (✉)
Department of Oceanography, US Naval Academy, Annapolis, MD, USA
e-mail: pguth@usna.edu

D. D. Scott
Adjunct Research Faculty, Colorado Mesa University, Grand Junction, CO, USA

P. L. Guth (ed.), *Military Geoscience*, Advances in Military Geosciences,
https://doi.org/10.1007/978-3-030-32173-4_3

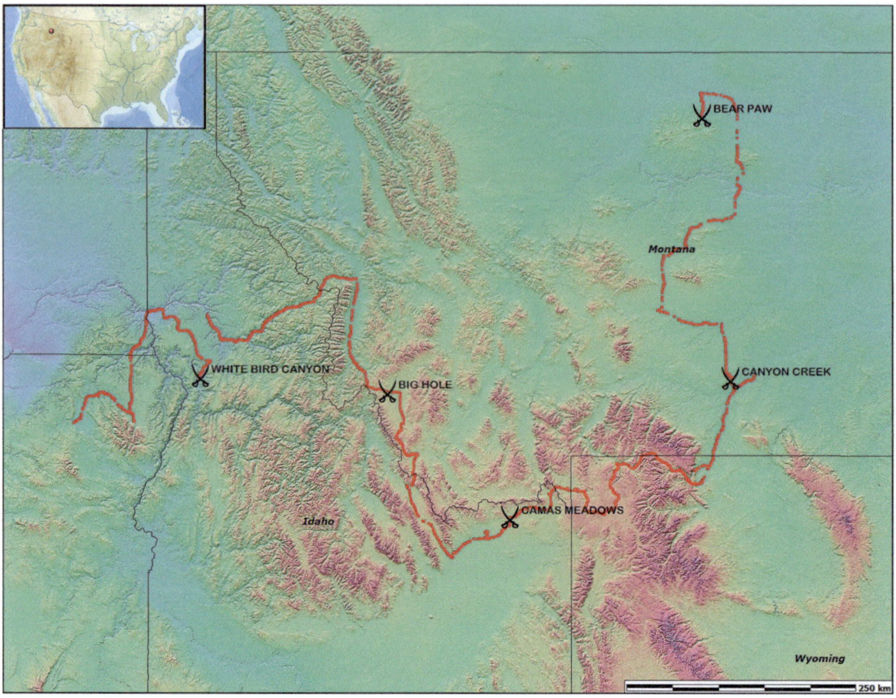

Fig. 3.1 Nez Perce War of 1877, with the Nez Perce trail and the major battles. (inset map from Wikipedia 2019)

The Battle of Big Hole

The Nez Perce War started in the spring of 1877 in southeastern Washington, northwestern Oregon, and adjacent Idaho when the US government attempted to move all Nez Perce bands onto a reservation. The Nez Perce won battles against the US Cavalry in Idaho at White Bird Canyon in June 17 and The Clearwater in July 11–12 and then attempted to move east into Montana where they naively thought they would be free. The US Army 7th Infantry fought the Nez Perce at the Battle of Big Hole on August 9–10, 1877. The battle had no winner, although it marked the first setback for the Nez Perce in their attempt to preserve their freedom, and they could not afford their losses in people and possessions.

Colonel John Gibbon collected the scattered companies of the 7th Infantry from garrisons in Montana, intercepted the trail of the Nez Perce south of Missoula, and overtook the Nez Perce where they were camped in a temporary village along the Big Hole River. The Nez Perce had about 200 warriors and 750 women and children under Chief Joseph. Gibbon attacked with 17 officers, 132 enlisted men, and 34 civilian volunteers, local civilians who augmented the military. Illustrative of the understrength frontier Army, the soldiers came from companies A, B, D, E, F, G, I, and K of the 7th Infantry as well as troop L of the 2nd Cavalry.

The Army lost 29 dead and 40 wounded; 6 soldiers eventually won the Congressional Medal of Honor. Of some 800 people in the Nez Perce camp, probably about 90 died, only a third of them warriors. After the battle, the Nez Perce continued east through Yellowstone National Park and then turned north toward Canada, fighting several more battles. Colonel Nelson A. Miles of the 5th Infantry, reinforced with units of the 2nd and 7th Cavalry totaling about 520 officers, men, and scouts, finally forced the Nez Perce to surrender in early October after the Battle of the Bear Paw (Fig. 3.1). This war created the reputation of Chief Joseph, one of the Nez Perce leaders.

The battle (Fig. 3.2) can be divided into the following phases (Scott 1994; Haines 1999; Greene 2000), which corresponds with the GIS and terrain analysis to be discussed later:

0. The Army approached the Nez Perce camp located on the east bank of the Big Hole River from the west, following a trail on the west side of the river at the base of the mountain (Fig. 3.2b, location 0). The slope just above this trail was generally grassy – this was where the Nez Perce grazed their horses – and the higher slopes were covered with dense pine forests. Between this trail and the village, the troops had to cross the marshy, meandering floodplain of the river that was covered with many small willow trees. Gibbon considered trying to drive off the Nez Perce horses before the assault but did not want to give up the element of surprise.
1. The initial attack on the Nez Perce camp took place before dawn and achieved complete tactical surprise (Fig. 3.2b, location 1). In line with Army doctrine of

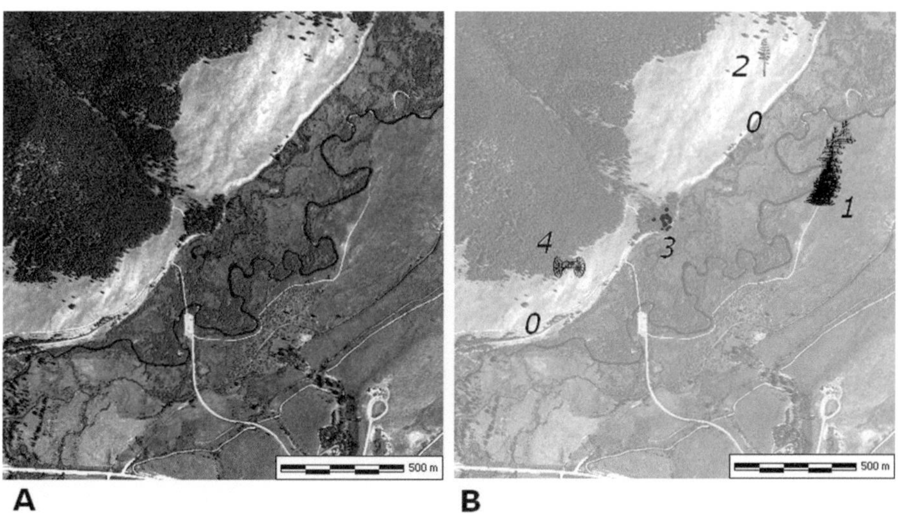

Fig. 3.2 (**a**) NAIP color aerial photograph of the Big Hole battlefield and (**b**) annotated version showing the key locations during the battle: (0) trail used by the Army to approach, (1) the Nez Perce camp, (2) location of the Nez Perce sniper in the Twin Trees on the hillslope with the horse herd, (3) the siege area, and (4) howitzer site

the time, Gibbon's force sought to destroy the teepees and other possessions of the Nez Perce to force them to submit to relocation on the reservation. Many of the Nez Perce casualties were women and children shot in the darkness with smoke from the weapons adding to the confusion of the battle. The soldiers attempted to burn the village, but the wet buffalo hide teepees would not easily ignite. This diversion of effort from attacking the Nez Perce to property destruction allowed the Nez Perce to regroup and attack the soldiers in the village.

2. Realizing the gravity of the situation, Colonel Gibbon ordered a retreat across the river to a forested alluvial fan on the west side of the valley (referred to from this point as the siege area). During the retreat, a Nez Perce sniper in the Twin Trees on the western hillslope harassed the soldiers until Army marksmen neutralized him (Fig. 3.2b, location 2).
3. Arriving at the siege area (Fig. 3.2b, location 3), Gibbon reasserted control over his forces and started them digging in. Two of the infantry companies had an experimental trowel bayonet (really an entrenching tool) that proved very effective.
4. As the soldiers arrived at the siege area, a 12-pounder mountain howitzer arrived on the hillside further west overlooking the battlefield and fired two rounds before the Nez Perce overran its position (Fig. 3.2b, location 4).
5. The soldiers spent the afternoon and evening of August 9 besieged on the alluvial fan (Fig. 3.2b, location 3). The 7th Infantry troops had provided the burial detail at Little Bighorn the year before, and there was concern about parallels between the two situations with soldiers surrounded and apparently outnumbered. The Nez Perce realized they could not overrun Gibbon's position without incurring prohibitive losses and thus attempted only to contain the soldiers while they tended to their dead before breaking camp. Gibbon was able to send messengers for help, and the next morning a courier announced that General Howard and reinforcements would arrive by noon from the northwest. The Nez Perce understood the meaning of the courier's arrival and retreated.

Methodology

We obtained digital data for the Big Hole battlefield from the US Geological Survey (USGS) and the state of Montana (Table 3.1). We took the AutoCAD files from the original archaeological work (Scott 1994) and converted them into shapefiles, a standard GIS format (ESRI 1998), using a combination of the freeware program fGIS (Brown 2005) and custom programming. A commercial mapping program supplied with the total station surveying instrument was used during the archaeology field work to create the original files, which were then manipulated in AutoCAD, and have several different internal structures. The original survey work used arbitrary coordinates, which we converted to latitude and longitude using the borders of the National Battlefield as recorded on the USGS Digital Raster Graphics and the Montana state cadastral files for Beaverhead County. The GIS allowed us to use and

Table 3.1 Digital data used

Data type	Data producer
Digital Elevation Model (DEM)	1/3 arc-second national elevation data from the USGS (lat/long coordinates)
Digital Raster Graphics, scanned topographic maps	1:24,000 scale, from the USGS (UTM coordinates)
National Agricultural Imagery Program (NAIP)	US Farm Services Agency, 1-meter resolution, natural color (state plane coordinates)
Digital Ortho Quarter Quads	USGS, 1-meter resolution, black and white (UTM coordinates)
Cadastral (land ownership)	Montana Cadastral Mapping Project, public-private partnership (state plane coordinates)

Table 3.2 Archaeological evidence of identifiable firearms

Firearm type (users)	Cartridge cases	Bullets	Minimum number of weapons
.44 Henry (Nez Perce)	26	14	10
Springfield M1873 .45/70 (soldiers)	524	131	90
Springfield M1868 .50/70 (volunteers)	38	41	24
15 other types	50	44	23
Total	638	230	147

After Scott (1994), p.78

merge data in UTM, state plane coordinate system, and latitude/longitude coordinates, and to plot the archaeological data on any of the base maps. All of the archaeological data files were changed into the standard shapefile format, which can be displayed in virtually any GIS software.

Scott (1994) estimated that the Army had about 154 Springfield rifles and carbines at the battle and that the Nez Perce could have had about 63 Springfields captured during battles earlier in the summer. Of the estimated total of 217 Army weapons used by both sides, he recovered evidence from at least 90 different weapons, or 41% of the estimated total. Based on the spatial pattern of the Springfield cartridges, Scott (1994) postulated that soldiers fired most of them. The volunteers used an older model Springfield with distinctive cartridges. Table 3.2 summarizes the recovered firearm evidence.

We used the MICRODEM freeware GIS program (Guth 1995, 2009) to create some customized analyses. Most of the work, however, could be done in any GIS program because all the data are in standard formats. We used the following GIS tools in our analysis:

1. Viewshed computations, which show the terrain visible from a particular location. The 10 m/third arc-second DEMs are accurate enough to show the general features of terrain blocking, but will not show microrelief such as banks along the river. Of greater concern, these viewsheds do not consider the vegetation. In general, digital vegetation data at a scale suitable for intervisibility studies does not exist, and the GIS algorithms have consequently not been developed to

incorporate such data. Work currently underway to include vegetation in MICRODEM's intervisibility algorithms has not been applied to the Big Hole battle analysis. Further, we would need vegetation data from 1877, and not what is present now, although, in general, we assume the vegetation was very similar in 1877 to what currently grows on the battlefield. The battlefield is now covered with three kinds of vegetation which affect the viewsheds differently: (1) grasslands, on the hillsides, which will not significantly affect intervisibility; (2) willows, along the river, that can locally affect intervisibility; and (3) pine forests, around the siege area and the howitzer site, which will greatly affect intervisibility. These vegetation types appear very differently in the color aerial photography, in terms of both color and texture, and by overlaying the viewsheds on the imagery, we can at least qualitatively discuss the effects of vegetation.

2. Animations showing the locations of individual weapons. The GIS can show the locations of all artifacts of a particular type, like the .45 caliber cartridges, and then highlight in another color each identified weapon and cycle through all the weapons of that type in turn. Unfortunately, whereas the graphical animation modes do not translate well to a printed report, they work well for analysis and visualization, display in an oral presentation, and on web pages.

Results

Figures 3.3 and 3.4 show the distribution of recovered Springfield and Henry cartridges, respectively, on the battlefield, and, in addition, Fig. 3.3 attempts to assign the Springfield cartridges to various phases of the battle (Table 3.3). Three phases of the battle overlapped in time and space, and we have not attempted to separate the initial attack and battle in the village from the retreat. We did attempt to separate the early Army offensive from the retreat to the siege area, but we suspect some

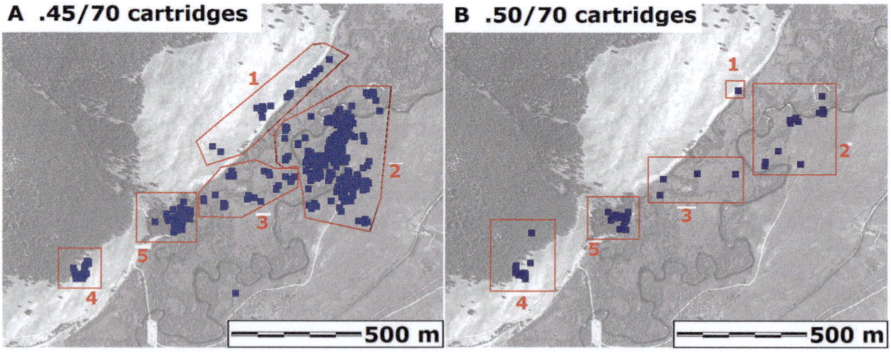

Fig. 3.3 Distribution of recovered Springfield cartridges (**a**) .45/70 used by soldiers and (**b**) .50/70 used by the volunteers. Numbers show phases of the battle: (1) initial volleys and covering fire, (2) fighting in the Nez Perce camp, (3) retreat, (4) howitzer arrival, and (5) siege area

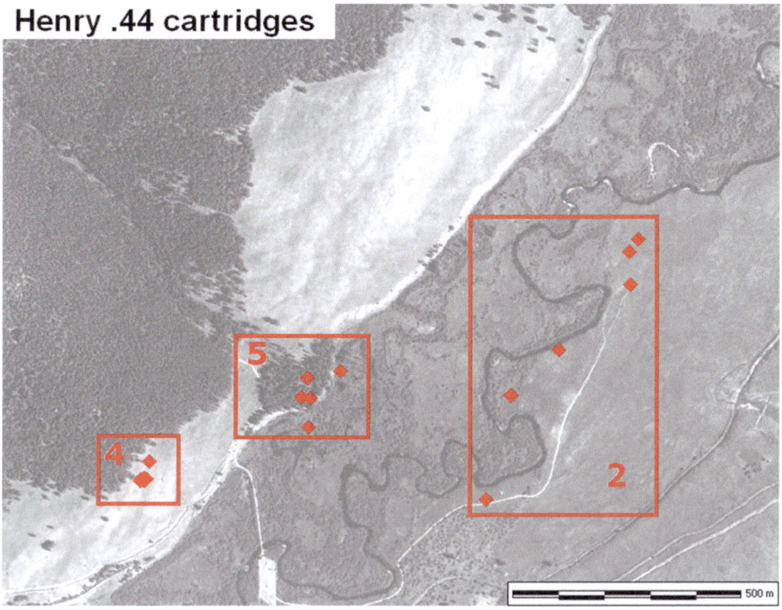

Fig. 3.4 Distribution of recovered .44 Henry cartridges on the battlefield, used exclusively by the Nez Perce. Numbered battle phases use the same key at Fig. 3.3. The Nez Perce were apparently not firing Henrys during the initial assault and the retreat from the village

Table 3.3 Assignment of identified cartridges with location data to battle phases

	.45/70	.50/70	Henry
Initial attack	320	17	10
Retreat	27	4	
Siege area	63	14	5
Howitzer area	20	16	8
Reserves	21	1	
Anomalous	1		
Total	452	53	23

overlap here. Because the Nez Perce had captured Army weapons, we used the locations of the Henry cartridges to get an idea of where undoubted Nez Perce warriors fired weapons during the battle because only the Nez Perce used Henrys. The Henry cartridges come from several distinctive groupings:

1. On the northeast margin of the village, where they overlap with a few Springfield cartridges
2. Inside a large meander bend southwest of the village, again with little overlap from the Springfields
3. Around the siege area, mostly outside the Army trenches
4. At the howitzer area

Table 3.4 Springfields with the most recovered cartridges

Weapon	Recovered cartridges	Used
M1873 35	21	Firing line northwest of village, retreat
M1873 43	18	Firing line northwest of village, inside village
M1873 39	15	Firing line northwest of village and one anomalous cartridge SW of the village
M1873 13	14	Firing line west of village, two in siege area
M1873 49	13	Firing line west of village, one cartridge south of village
M1873 32	11	Firing line northwest of village, east side of village
M1873 25	11	Howitzer area
M1873 29	9	Firing lines, retreat, and siege area
M1873 22	9	Reserve and siege area
M1873 16	9	Firing lines, retreat, and siege area
M1873 6	9	Retreat
M1868 S-15	9	Firing line north of village, two in siege area

Table 3.4 shows the Springfields with the most recovered cartridges, and Fig. 3.4 comprises maps showing the locations of each of these weapons. These 12 weapons (11% of those identified) accounted for 148 (29%) of the recovered cartridges. None of the weapons have a pattern consistent with use by a Nez Perce warrior. Cartridge 1355, a .45/70 fired from Springfield 39, occurs 530 m southwest of the village. The only cartridge close to this location (160 m to the east) comes from a Henry. These two positions are more likely Nez Perce than Army. Springfield 39 fired 14 recovered cartridges on the firing lines northwest of the village, but none in the retreat or siege. We suggest that this weapon was lost by a US soldier in the initial attack and was found and later fired by a Nez Perce on the extreme southern margin of the battlefield.

Figure 3.5 compares the patterns of .45/70 and .50/70 cartridges near the village and suggests that the volunteers armed with the M1868 did not play nearly as significant role in the battle as the regular Army infantry troops. They fired many fewer rounds, and the village yielded only one .50/70 cartridge. The 34 volunteers represented 18.5% of the total force, but their cartridges represent only 10.5% of those recovered. Army Officer Lt. James Bradley had been placed in charge of the volunteers on the north edge of the Army formation, but he was killed at the start of the battle which may have contributed to the ineffectiveness of the volunteers, although they may never have been capable of the discipline displayed by the professional soldiers. If Nez Perce warriors fired the 16 .50/70 cartridges in the howitzer area, the volunteers fired only 7.6% of the Army shots (Fig. 3.6).

One of the Henry cartridges found SW of the village occurs in close proximity to four .45/70 cartridges fired from Springfields 53 and 55 (Fig. 3.7). The overall pattern from these two weapons, fired south and east of the village and from a location on the hillslope NE of the siege area, suggests they could only have been used by Nez Perce warriors throughout the battle. The .45 bullets found in the same area

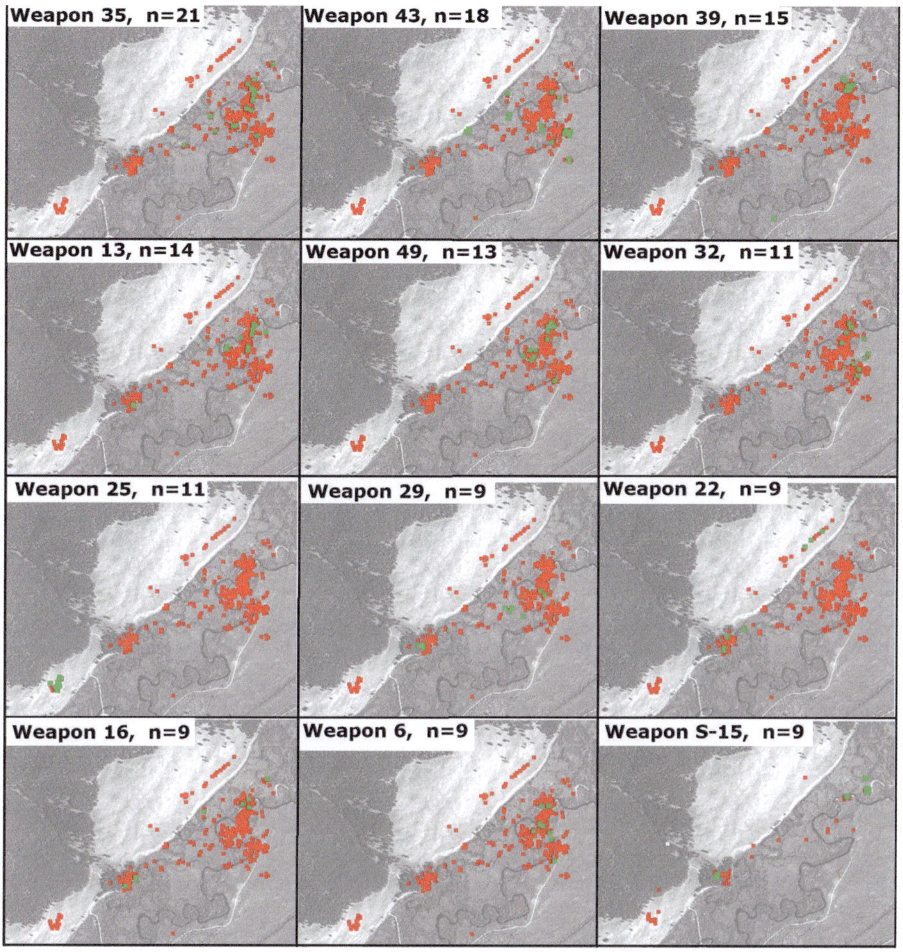

Fig. 3.5 Springfield rifles each with at least nine recovered and identified cartridges. Red squares show all cartridges for that type, and green squares highlight particular weapons. Weapons 35, 43, 39, 13, 49, 32, 25, 29, 22, 16, and 6 are M1873s firing the .45/70 cartridge used by soldiers, and weapon S-15 is an M1868 firing the .50/70 cartridge used by the volunteers who attacked from the northernmost position

suggest that the Army was firing on this location, supporting the suggestion that it was a Nez Perce position. Weapon 43 (Fig. 3.4) has an interesting pattern that suggests it may also have been used by a Nez Perce warrior rather than a soldier. It seems to circle the periphery of the village, always near the outer limits where cartridges have been recovered. Weapon 80 fired six rounds in a line from the village, west to a point on the hillside overlooking the battlefield (Fig. 3.6), where it is associated with a cartridge fired from Weapon 53. The weapons fans computed for these rifles (Fig. 3.8) show they could not have reached the Twin Trees. They do

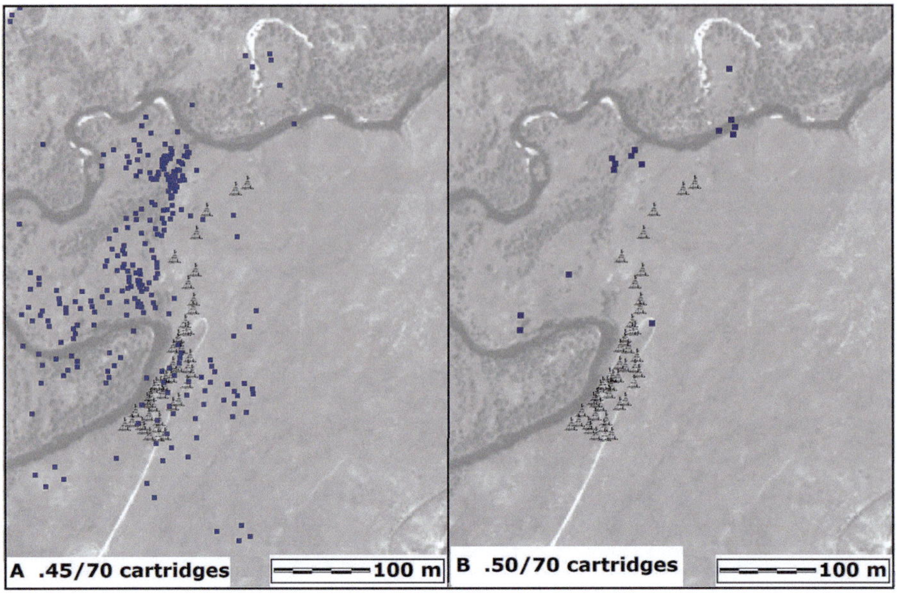

Fig. 3.6 Distribution of recovered Springfield cartridges near the Nez Perce village (**a**) .45/70 used by soldiers and (**b**) .50/70 used by the volunteers

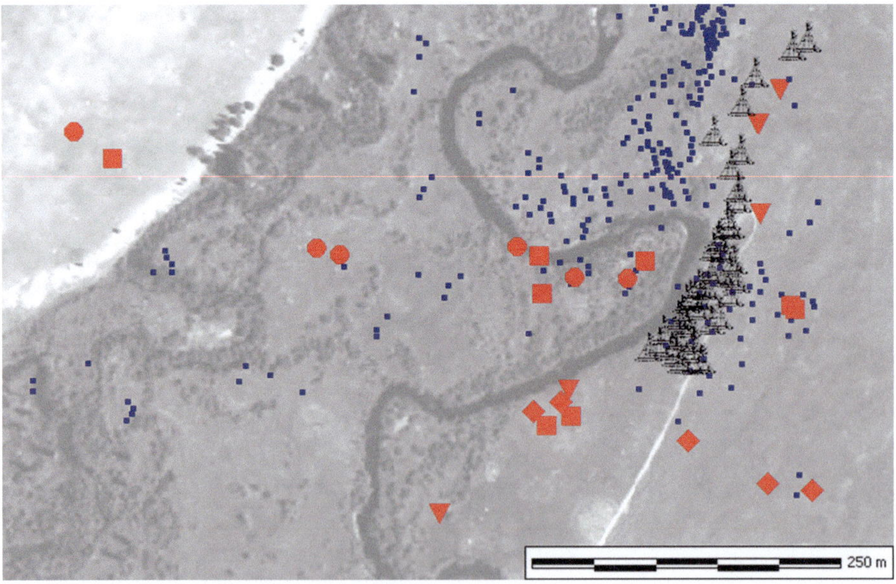

Fig. 3.7 Recovered cartridges from Springfields 53 (squares), 55 (diamonds), and 80 (circles) and Henrys (triangles), all shown in red, which may have been used by Nez Perce warriors. Blue squares are Army cartridges

cover the village, the retreat, and the siege area, and given their lack of participation in the firing lines northwest of the village and their association with other weapons likely used by the Nez Perce, this suggests that the cartridges from this location on the hill were fired on the US infantry soldiers.

The 14 Springfield cartridges recovered inside an abandoned meander north of the village (Fig. 3.6) initially suggested a Nez Perce fighting position. They came from six identified weapons (Table 3.5), five of which also fired a total of 22 rounds on the Army firing lines NW of the village. Two additional weapons also fired six rounds from inside the siege area. The overall pattern suggests a small US Army squad crossed the river and went about 130 m north of the northernmost teepees in the village, where it was involved in a fierce firefight. The squad included two volunteers (armed with .50/70 Springfields S-15 and S-16) and at least four 7th Infantry troops.

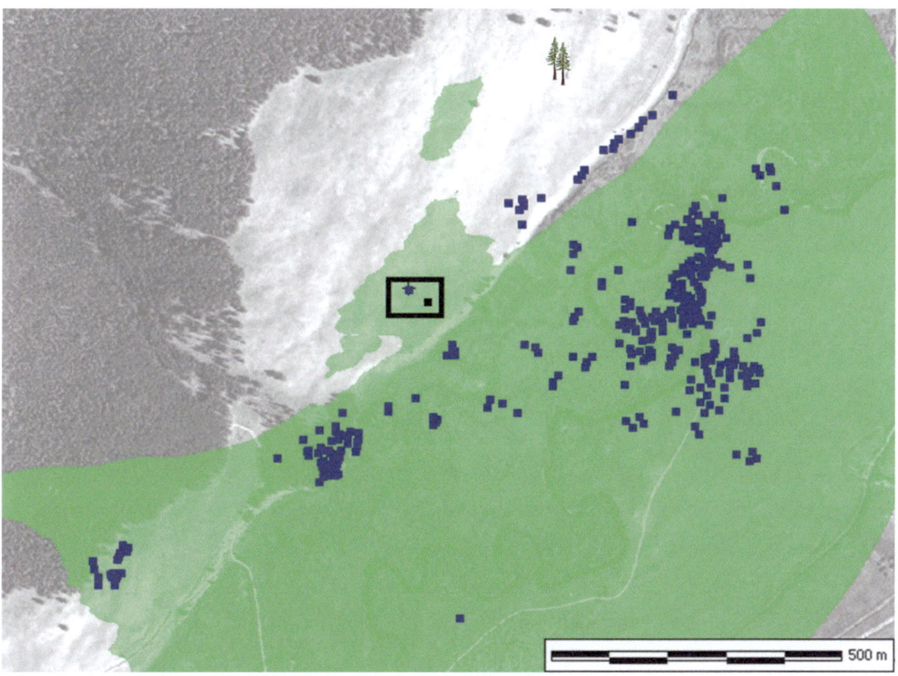

Fig. 3.8 Weapons fan (green) for Cartridge 1644 fired from Weapon 80 (black square), with a 1000 yard range and both weapon and target 1 m above the ground. The fan for Cartridge 1646 from Weapon 53, recovered just down the hillslope, covers essentially the same area. Twin Trees out of sight northeast of this position

Table 3.5 Weapons used inside the abandoned meander NE of the village

Weapon	Cartridges in meander	Other cartridges from weapon
S-15	6	1 on firing lines NW of village 2 in siege area
S-16	1	
16	1	3 on firing lines NW of village 1 during retreat or very early in attack 4 in siege area
35	1	10 on firing lines NW of village 1 at south end of village 5 during retreat
47	2	4 on firing lines NW of village 1 during retreat or very early in attack
88	1	4 on firing lines NW of village
Unknown	2	

Sniper in Twin Trees

Histories of the battle emphasize the importance of a Nez Perce sniper in the Twin Trees on the hillslope northwest of the village, and Colonel Gibbons ordered his troops to neutralize the sniper. Gibbons' description suggests some hyperbole about the sniper's effectiveness, which could have been most effective in reducing Army morale. Figure 3.9 shows the weapons fan from this location, which depends very little on how high up in the tree the marksman was. Scott (1994) recovered five .45-405 bullets fired from Army Springfields near Twin Trees with nose orientations suggesting a firing azimuth of about 330° and whose nose impact damage suggested they were fired at medium range (500–900 yards). The pattern suggests that an Army marksman walked his fire up the hillside to find the correct range and the smoke from the Springfield would blind the shooter and require a spotter to see where the rounds hit.

Figure 3.10 shows the locations of these bullets and a geometric construction showing potential locations for the Army counter-sniper fire. The sectors outlined in red show cartridges with the correct range, and an azimuth within 20° of the bullet nose orientations; 118 of the 452 recovered cartridges were found in the potential zone, and 41 of the 87 identified weapons. Table 3.6 shows the most likely weapons used, assuming that it would have taken several rounds to adjust the range to hit the sniper. The reserves along the base of the hill could not have fired the shots, as they were too close or were positioned so that the bullets they fired would not have the same orientations as those identified by Scott (1994). Figure 3.10 shows the most likely weapons used against the sniper, assuming that a number of shots would have been required from a single location and that preservational bias in the recovered cartridges does not significantly affect the results. The most likely locations for this fire appear to be the west side of the village, or from within the large meander just to the west of the village.

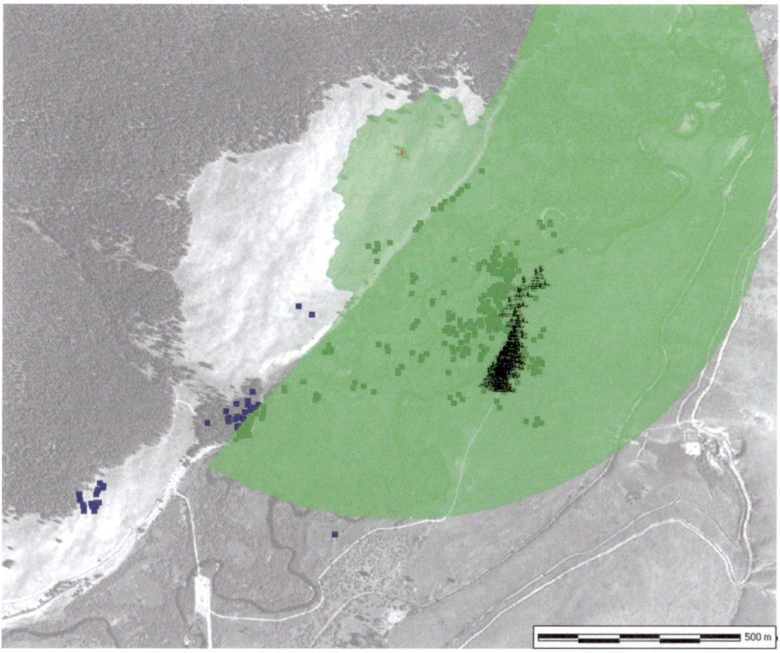

Fig. 3.9 Weapons fan (green) from a sniper in the Twin Trees. This is for a position 5 m above the ground, but the village and retreat route are within the weapons fan even if firing occurred at ground level. The siege area is not covered by the weapons fan from the Twin Trees from any reasonable elevation

Mountain Howitzer

Colonel Gibbon had been an artillery officer before the American Civil War (1861–65), and prior to the Civil War, he had written a book that became one of the key artillery manuals during that war (Gibbon 1860). At the Battle of Big Hole, he had a 12-pounder mountain howitzer, but fearing the noise it would make approaching through the dense underbrush, he had left it along the route with instructions to arrive later. Lacking trained artillery troops, two infantry sergeants and four soldiers manned the gun.

Figure 3.11 shows the weapons fan for the howitzer's position based on its maximum range of 1000 yards. The Nez Perce village was 150–350 yards beyond the maximum range. The extreme altitude of the weapon, 70 m vertically above the village, may have increased the effective range enough to reach the southern end of the village, where two shells have been recovered (Scott 1994). The howitzer was well placed to cover the retreat from the village to the siege area.

The weapons fan neglects the effect of vegetation, but the trees would likely have had negligible impact on the results. The small willows along the river would not affect artillery, and from the howitzer's position, the ponderosa pine forest on the

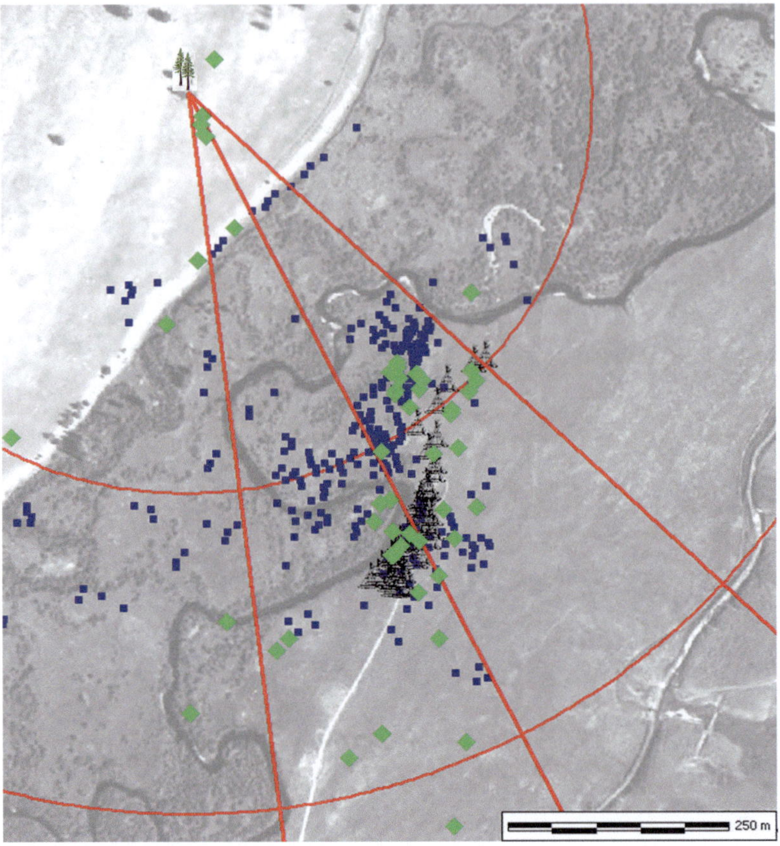

Fig. 3.10 The .45 bullets recovered on the battlefield (green diamonds) and the .45/70 cartridges (small blue squares) from the Army Springfield rifles. Red circles outline the medium range suggested by bullet nose impact, and the radial lines show the suggested 330° firing azimuth with a 20° window

higher slopes is almost entirely masked by topography. Had the Nez Perce been able to work the weapon after its capture (Haines (1999) notes that some of the Nez Perce may have seen artillery during earlier campaigns), the ponderosa pines in the siege area could have enhanced the effect of exploding shells.

The weapons fan in Fig. 3.12 underscores the vulnerability to attack of the howitzer's position from the south or southwest. In those directions, the terrain blocks visibility until attackers reach about 100 m from the gun. After firing two rounds (one of which was apparently an exploding shell which was not fused and did not explode), the Nez Perce overran the howitzer and killed one of the six crew members. In the heat of battle, the howitzer crew could not see approaching Nez Perce until the last minute. Cartridges surrounding the howitzer document this fight and reflect at least 23 different weapons (Table 3.7). The .45/70 cartridges were probably fired by the soldiers and the other two types by the Nez Perce.

Table 3.6 Springfields in position to fire against Twin Trees sniper

Weapons	Cartridges in zone	Notes
43	10	Nez Perce weapon?
53	7	Nez Perce weapon?
32	6	No clusters
33	6	6 in small area of village
55	5	Nez Perce weapon?
6	5	No clusters
2	4	No clusters
28	4	3 in meander
35	4	2 in meander
7	4	4 in village
13	3	2 in willows
29	3	2 in willows
30	3	No clusters
48	3	3 in village
49	3	2 in willows
64	3	No clusters
80	3	Nez Perce weapon?

The six soldiers assigned to the howitzer would have been armed with the M1873 .45/70 Springfields, and Scott (1994) identified cartridges from four M1873 rifles. Given that the soldiers were trying to fire the howitzer and deal with its mule team, they might not all have had time to use their rifles. Scott (1994) counted 11 cartridges from Weapon 25, 4 from Weapon 23, and 2 each from weapons 24 and 26. In addition to the six-man gun crew, two civilians, an officer's servant, and a guide bringing up a mule with 2000 rounds of ammunition for the .45/70 Springfields were probably also with the howitzer (Haines 1999), but it is unclear how they might have been armed.

Haines (1999) noted a string of four .45/70 cartridges NE of the howitzer and attributes them to a Nez Perce who attacked the howitzer. Scott (1994) identified 11 rounds fired by that weapon, number 25, around the howitzer position. Number 25 was one of the most prolific Springfields, as only five weapons had more recovered cartridges; one other also had 11 (Fig. 3.5 and Table 3.4).

In contrast to the M1873 Springfields which fired a large number of cartridges from a few weapons, the 11 identified M1868 Springfields fired only 13 recovered .50/70 cartridges, with weapons S-5 and S-9 firing two recovered cartridges each. An additional .50/70 single cartridge was fired from a Sharps rifle. Only two of the 1868 Springfields used in the howitzer area had a second cartridge recovered elsewhere on the battlefield (S-6 with an anomalous cartridge about 150 m to the north in dense forest and S-22 in the willows between the village and siege area). Since apart from the servant and guide there is no evidence of volunteers armed with the older Springfields in the vicinity of the howitzer, the Nez Perce probably used these weapons.

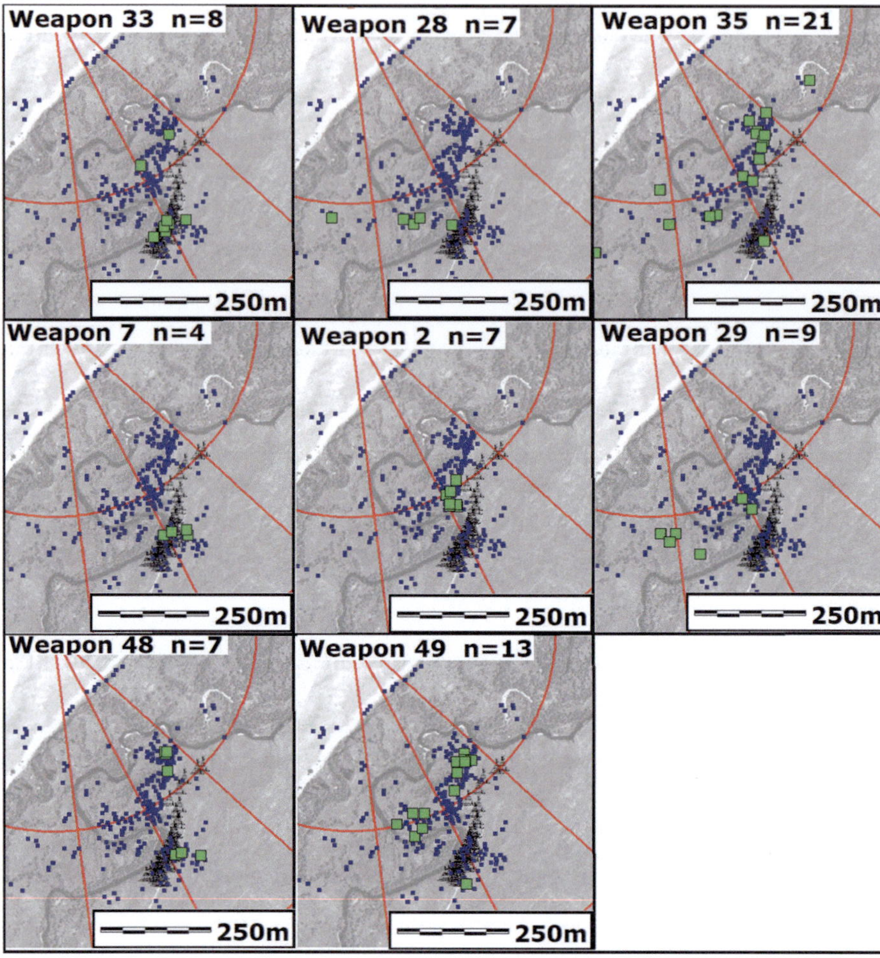

Fig. 3.11 Candidate weapons used for the anti-sniper fire (green squares are recovered cartridges from single weapon; blue squares are all Army cartridges)

Discussion

Metal detector surveying at the Big Hole Battlefield recovered a large number of metal cartridges and bullets. The use of different models and calibers of weapons allows reasonable differentiation of the three groups of combatants (Nez Perce, soldiers, and civilian volunteers armed with an older model of the Springfield), although the Nez Perce may have captured and reused Army weapons during earlier encounters and during the early stages of the Big Hole battle. Forensic analysis of the casing allows identification of individual weapons; plotting their locations on a map shows the movement of the weapon during the battle. Knowing the timeline of

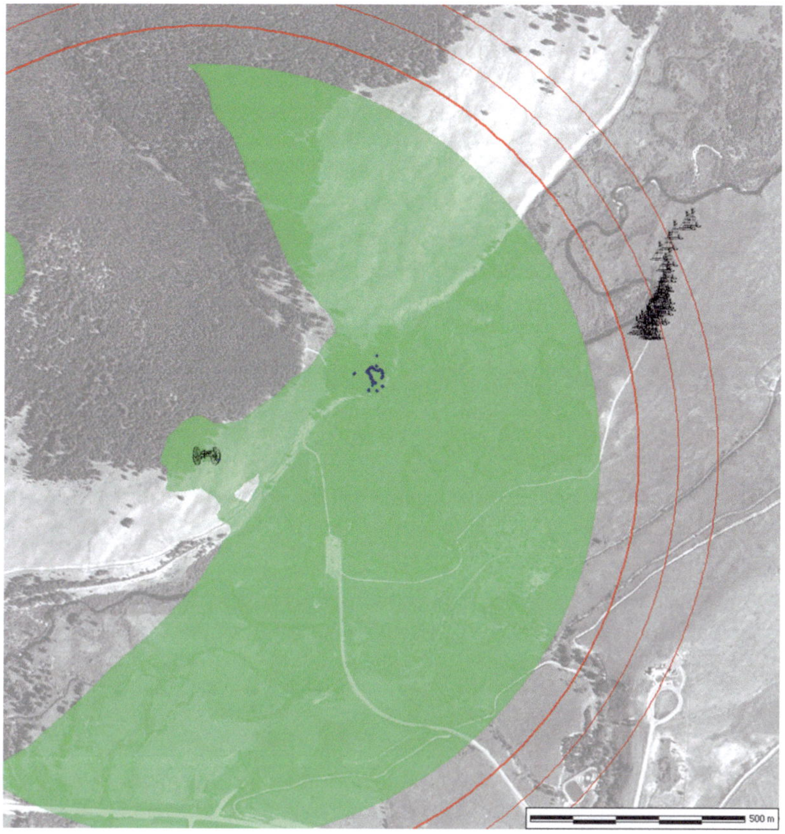

Fig. 3.12 Weapons fan for the mountain howitzer, disregarding vegetation. The howitzer had a range of 1000 yards, and the red arcs to the right of the fan show 1100, 1200, and 1300 yard ranges. The rifle pits (blue) outline the siege area about 400 m NE of the howitzer's position

Table 3.7 Cartridges found in the howitzer area

Weapon type	Cartridges	Number of weapons	Unidentified
.45/70 Springfields	20	4	1
.50/70 Springfield (15) and Sharps (1)	16	12	2
.44 Henrys	9	5	
Total	44	23	

the battle allows making a reasonable path for each weapon. While it is tempting to attribute those paths to a single combatant, someone could have picked up a discarded weapon and used it later in the battle. Despite these caveats, the patterns of recovered cartridges show the phases of the battle (Figs. 3.3 and 3.4). The dense concentrations of Army cartridges to the northwest of the Nez Perce camp show the discipline of the Army troops, in contrast to many fewer rounds fired by the volunteers. Many fewer distinctive Henry repeating rifle cartridges were recovered; their locations were concentrated on the capture of the howitzer and around the siege area (Fig. 3.4). The Nez Perce were forced to use captured Army weapons to supplement their limited number of repeating rifles; Table 3.6 shows four Army Springfields that were probably used by the Nez Perce based on their recovery locations.

Viewshed analyses show the impact of a Nez Perce sniper in a tree on the hillslope overlooking the battlefield. Combined with bullets found downhill from the tree, which supply an estimated range and firing azimuth, the analysis suggests which Army weapon engaged and neutralized the sniper. This brings to life Gibbons' account of the mood swing among the Army troops as the sniper first attacked the retreating soldiers, and then relief when the firing ended. The viewsheds also show the ineffective use of the mountain howitzer, employed too far from the village, and then quickly overrun because the steep topography blocked the view of the approaches.

GIS reveals specific details of the movements during the battle, what was visible and masked to participants on the ground, and complements the written record. While Big Hole was a small battle, it played a decisive role in the campaign. GIS allows rapid, interactive cycling through the data on screen to explore the data prior to picking static maps such as shown in this paper.

Conclusion

Reanalysis of the archaeological data collected at the Big Hole Battlefield using imagery and analysis techniques available in a GIS allowed new insight into the battle. Plotting artifacts on a map or image background provided much greater context than would be possible just using drafting software, and the ability of GIS to rapidly and easily filter the data allowed rapid creation of maps showing each weapon identified through ballistic analysis of the cartridges. Viewshed analysis applied to the employment of weapons shows how the battle unfolded, especially the multiple errors in positioning the mountain howitzer and why it was so easily overrun. Big Hole was a small but fierce battle, and its limited scope and the recovery of a high percentage of the cartridges and bullets fired during the battle have allowed a unique, objective analysis of the battle itself.

References

Brown, B. (2005). fGIS Program and help file: University of Wisconsin and Wisconsin Department of Natural Resources, available at http://www.forestpal.com/fgis.html. Last accessed 18 June 2019, but program has not been updated since 2005.

ESRI. (1998). ESRI shapefile technical description: ESRI White paper. Available online at http://www.esri.com/library/whitepapers/pdfs/shapefile.pdf. Last accessed 18 June 2019.

Fox, R. A., Jr. (2003). *Archaeology, history, and Custer's last battle: The Little Big Horn reexamined.* Norman: University of Oklahoma Press.

Gibbon, J. (1860). *The artillerist's manual, compiled from various sources, and adapted to the service of the United States*. New York: D.Van Nostrand.

Greene, J. A. (2000). *Nez Perce summer 1877: The U.S. Army and the Nee-Me-Poo crisis*. Helena: Montana Historical Society Press.

Guth, P. L. (1995). Slope and aspect calculations on gridded digital elevation models: Examples from a geomorphometric toolbox for personal computers. *Zeitschrift für Geomorphologie N.F. Supplementband, 101*, 31–52.

Guth, P. L. (2009). Geomorphometry in MICRODEM. In T. Hengl & H. I. Reuter (Eds.), *Geomorphometry: Concepts, software, applications. Developments in soil science series* (pp. 351–366). Amsterdam, Elsevier.

Haines, A. L. (1999). *An elusive victory: The battle of the Big Hole*. Helena: Falcon Publishing.

National Park Service. (2016). *Big Hole*, https://www.nps.gov/biho/planyourvisit/upload/FINAL-VERSION_-BIHO-unigrid_Side-2_X3_-compressed-for-web.pdf. Last accessed 18 June 2019.

Scott, D. D. (1994). *A sharp little affair: The archeology of the Big Hole Battlefield*. Reprints in Anthropology. Lincoln: J & L Reprint Company.

Scott, D. D., Fox, R. A., Jr., Connor, M. A., & Harmon, D. (2000). *Archaeological perspectives on the battle of Little Bighorn* (New ed.). Norman: University of Oklahoma Press.

Wikipedia. (2019). Big Hole National Battlefield, https://en.wikipedia.org/wiki/Big_Hole_National_Battlefield. Last accessed 19 June 2019.

Chapter 4
The Influence of Physical Geography on the Outcome of the Battle of Spioenkop During the Anglo-Boer War, 1899–1902

H. A. P. Smit and J. Bezuidenhout

Military Geoscientists and the Influence of Geographical Factors on Battles

> Physical geography is literally inescapable (Colin S. Gray (1999), p. 163).

Military geographers have a long tradition of studying battles by analyzing the physical geographical factors that influenced the outcome of battles (Palka 2011a, b, c). Modern geoscientists continue this trend and continuously expand the corpus of literature dealing with this important aspect of military geoscientific investigation. A diversity of geographical aspects were investigated by authors such as Doyle and Bennet (1999), who analyzed how terrain influenced the outcome of the Gallipoli Campaign in 1915; Puckett (1994) and Galgano (2011), who studied the role of streams in military operations; Müller (2006), who identified geographic key factors in global operations; Tate (2006) who explained how terrain analysis can aid military decision-making; Byers and Guth (2009), Gnaser (2009), and Mang (2009), who discussed the effect of mountains on military operations; Winters, Galloway, Reynolds, and Rhyne (1998), who brilliantly explained the effect of weather and climate on battles; and Palka (2011, b, c), who analyzed both the physical and human settings and their influence on military operations in Afghanistan and Iraq. Furthermore, entire books have been dedicated to specific geographical aspects and their influence on military operations. Two such books are *Battling the Elements: Weather and Terrain in the Conduct of War* (Winters et al. 1998) and *Fields of Battle: Terrain in Military History* (Doyle and Bennett 2010). In research like this,

H. A. P. Smit (✉) · J. Bezuidenhout
Department of Military Geography & Department of Physics, Faculty of Military Science, Stellenbosch University, Stellenbosch, South Africa

Military Academy, Saldanha, South Africa
e-mail: hennies@ma2.sun.ac.za

P. L. Guth (ed.), *Military Geoscience*, Advances in Military Geosciences,
https://doi.org/10.1007/978-3-030-32173-4_4

the influence of geographical factors on the outcomes of battles is analyzed and discussed.

In a South African context, authors like Ackermann (1957), Opperman (1981), Perry (1996), Pretorius (2002), Scholtz (2005), Smit (2011), and Van der Waag (2013) are only some examples of researchers writing about the influence of geographical factors in South African military operations. In the next two sections, a brief summary of the Anglo-Boer War and of the battle of Spioenkop will provide the backdrop of an analysis and discussion on the physical geography that influenced the outcome of this battle.

The Anglo-Boer War

The Anglo-Boer War raged between Britain and its Colonies on the one side and the two Boer Republics, the Zuid-Afrikaansche Republiek (ZAR), also called Transvaal, and, its ally, the Orange Free State (OFS), from 11 October 1899 to 31 May 1902. What British intelligence initially perceived as a small war that would last a few months turned out to be a protracted conflict lasting 2.5 years, mainly because the British underestimated the Boer Republics and the influence of the geography of South Africa on armed conflict (Sleigh 1979; Wessels 1991).

Initially the British plan was to concentrate its forces in Cape Town and then use the railroad as line of advance to first attack the Free State and then invade the Transvaal via the Free State capital Bloemfontein. Using the railway line in this way would also ensure the adequate resupply of the attacking force, as the railway line allowed for the quick movement of the huge amount of supplies needed for such an undertaking. The falling of Pretoria, the ZAR capital, would signal the end of the war, since taking possession of the capital city was the customary way in which to defeat an enemy during that time (Scholtz 2000).

At 17h00 on Wednesday, 11 October 1899, the ultimatum delivered to the British government by the Boer Republics expired, starting the Anglo-Boer War. The Boers moved quickly and invaded the British Colonies, laying siege to British troops in various towns. In Natal the Boers won a series of battles and trapped General George White and 14,500 men in Ladysmith. This was a serious blow to the British plans, and General Sir Redvers Buller, commander of the British forces in South Africa, had to go back to the drawing board and rethink his plan of attack (Ransford 1969).

The priority now shifted to relieving the trapped troops in the various towns, especially those in Kimberley, where one of the most prominent British politicians, Cecil John Rhodes, was trapped, and Ladysmith, where a large part of the British fighting force were surrounded (see Fig. 4.1 below). While Buller was preparing his forces for this new plan of attack, the Boers entrenched themselves in the foothills of the Drakensberg, along the northern banks of the Tugela River.

Buller's new plan entailed taking 21,600 men to Natal to relieve Sir George White in Ladysmith and sending General Methuen with 15,000 men to deal with the

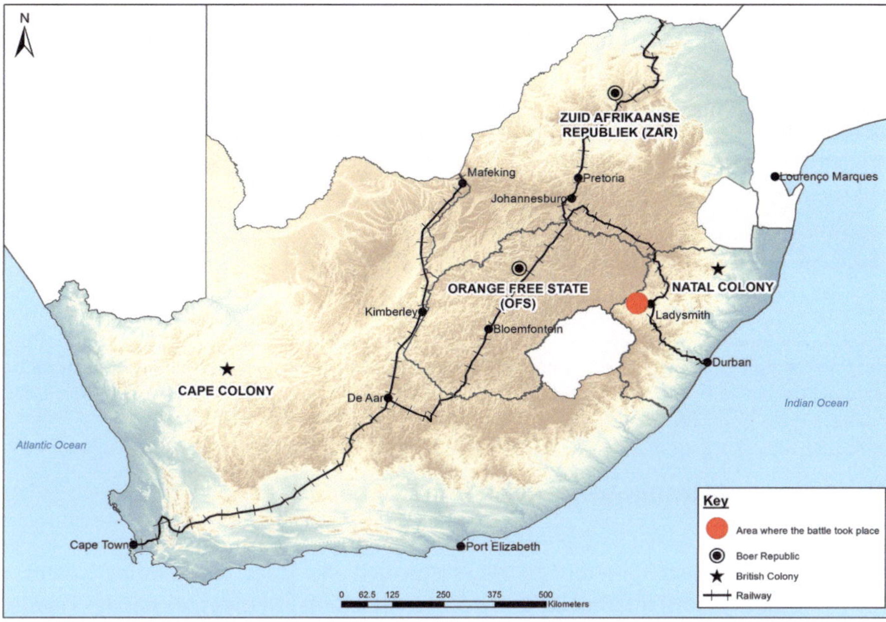

Fig. 4.1 Map of the British Colonies and Boer Republics during the Anglo-Boer War

situation in Kimberley (Pakenham 1979). On 15 December 1899, Buller attacked the formidable defensive line at Colenso that stood between him and the trapped General White in Ladysmith and was driven back with heavy casualties. This defeat led to his replacement as commander-in-chief of the British forces in South Africa by Field Marshal Lord Roberts. Buller retained command of the forces in Natal, and by January 1900 he was ready to try and cross the Tugela again (Pakenham 1979).

On 24 January 1900, the second bloodiest battle of the Anglo-Boer War was fought at Spioenkop, a high hill overlooking the Tugela River in the eastern part of present-day South Africa. On that day 58 Boer soldiers lost their lives, while the British causalities were 225 men killed, 550 wounded, and more than 300 taken prisoner, or missing in action (Breytenbach 1973; see Fig. 4.2).

While most military conflicts occur against the backdrop of the geography of the surrounding area, at Spioenkop, physical geography took center stage from the start. A series of bad decisions further enhanced the importance of geographical factors and ultimately led to the massive British losses and the Boer victory. Although the number of casualties sustained on the day might seem insignificant when compared to losses during later conflicts like the First World War, it was staggering in the context of the battles fought during the Anglo-Boer War.

Fig. 4.2 Photo **a**: Spioenkop as seen from the summit toward the Tugela. Note the effect of not being on the true crest line – the whole area beyond the crest line is dead ground. The dam in the Tugela in the background was only built long after the battle. Photo **b**: Part of the main trench that became a mass grave for the fallen British soldiers. (Photos: J. Bezuidenhout)

The Battle of Spioenkop

After the unsuccessful attempt to break through the Boer defenses at Colenso, Buller decided to outflank Spioenkop, a 530-meter-high hill overlooking the Tugela. He hoped that this would force the Boers to abandon the hill, opening a gap for him to force a way through the Boer defenses. The task to execute the flanking move fell on Lieutenant General Sir Charles Warren, with a 1700 strong force under him. On Monday, 22 January 1901, after a meeting with Buller, Warren decided that the only viable course of action would be to attack and hold Spioenkop itself (Breytenbach 1973).

Initially everything went according to plan. The attack on the main summit was led by Major General E.R.P. Woodgate. With intermittent light rain falling, Woodgate and his force climbed Spioenkop and brushed aside the small contingent of Boers on the hill. Then they attempted to dig trenches in the shallow, stony ground. Thick fog obscured their surroundings; they had no adequate maps of the area and had failed to conduct proper reconnaissance of Spioenkop (Amery, 1902; Ransford 1969; Breytenbach 1973; Pakenham 1979; Sleigh 1979). When the fog started to lift, Woodgate realized that they were in a dangerous situation. Not only were the trenches shallow, more importantly, they were in the wrong place. Woodgate had mistaken a slight undulation on the crest as the crest line and had ordered his soldiers to dig in there. They were in fact several meters from the crest line, giving the Boers ample dead ground to launch a counterattack (Amery 1902; Scholtz 2005). What made a potentially dangerous situation a deathtrap was the fact that Spioenkop was not the natural fortress the British thought it was. The trenches were overlooked by a Boer-held hill called Aloe Knoll 1.8 km to the east. From there, the Boers poured rifle fire and artillery shells into the heart of the British trenches. At the same time, the Boers were sheltered from British artillery fire from opposite the Tugela (Amery 1902; Smurthwaite 1999; Dye 2012; see Fig. 4.3).

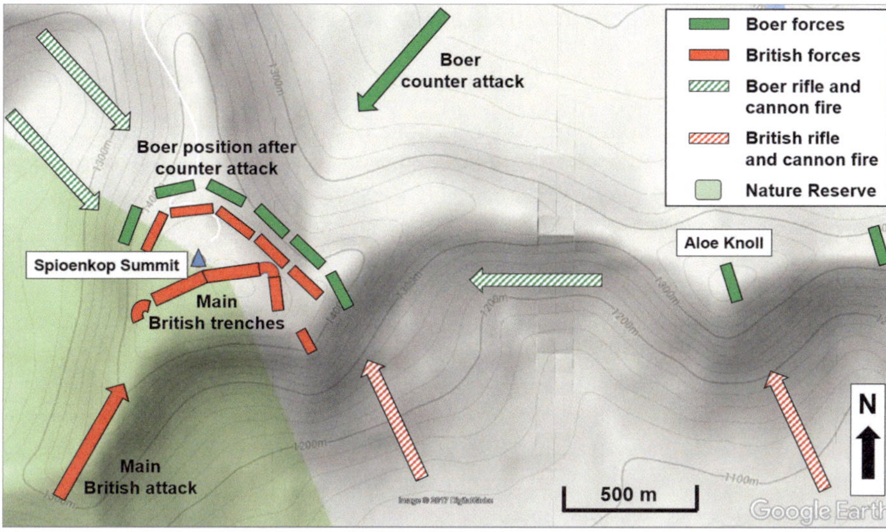

Fig. 4.3 Google Earth Map of the battle of Spioenkop, indicating the firing lines and trenches of the Boer attackers and the British defenders

Later in the day, the Boers counterattacked and stormed Spioenkop. Woodgate and his soldiers were trapped but fought bravely. By nightfall the British retreated. Ironically, the Boers were doing the same. Although General Botha, the senior officer in charge of the Boer forces, thought that the British would retreat during the night, most of the Boer soldiers felt they had enough and abandoned their positions. Only two Boer soldiers that witnessed the British retreat allowed Botha to order his troops up Spioenkop again, giving this vital hill back to the Boers.

The Geographical Factors that Influenced the Battle

Four geographical factors played a major role in the outcome of the battle, lack of adequate terrain intelligence and maps that hindered effective reconnaissance, the thick fog that enveloped the battlefield during the British ascent of Spioenkop, the rain and consequent pitch dark night that made the ascent extremely difficult, and the inability of the British artillery to engage the Boer artillery because of the terrain. These factors will be discussed in more detail below.

Lack of Adequate Terrain Intelligence and Maps Constricted Reconnaissance

Intelligence of a battlefield forms the backbone of any battle plan, but this intelligence regarding Spioenkop was sorely lacking on the British side. According to Jacobs and Smit (2004), the British had only poor maps at their disposal. Ransford (1969) concurred with this conclusion of Jacobs and Smit (2004). Pakenham (1979) stated that Buller had to rely on a survey done by the Field Intelligence Department and a microfilmed map that arrived from Ladysmith via carrier pigeon. The map legend stated "vacant spaces indicate that data is wanting, rather than that the ground is flat" (Pakenham 1979, p. 208). Breytenbach (1973) stressed Buller's dilemma and indicated that a vital prerequisite for a successful attack on Spioenkop was a thorough knowledge of the summit of Spioenkop and the surrounding area. To this end, the maps at Buller's disposal were inadequate. The mountainous terrain and the fact that the Boers dominated the high ground around Spioenkop made standard reconnaissance virtually impossible, and consequently Woodgate's troops arrived on the summit of Spioenkop without adequate knowledge of the position they had to defend against a Boer attempt to retake the Kop. Virtually all the sources analyzed (Ackermann 1957; Amery 1902; Ransford 1969; Breytenbach 1973; Pakenham 1979) indicated that this was a serious weakness of the British plan. Although the British had a balloon at their disposal from which they could easily have compiled a sketch map of the Spioenkop summit, it was not done. This caused the British to entrench in exposed positions, with disastrous consequences.

The Thick Fog that Obscured the True Nature of the Battlefield from the British

To further complicate matters for the British, a thick fog covered Spioenkop when the British soldiers reached the summit (Amery 1902; Farwell 1977; Kestell 1999). According to Breytenbach (1973), visibility was limited to a few paces. The almost total lack of battlefield intelligence, even after reaching their objective, caused General Woodgate to give the command to dig north-facing trenches. From these trenches, he assumed the ground fell steadily away in a northerly direction, the direction from which he expected the Boer attack. Amery (1902) criticized Woodgate for not sending out scouts to accurately determine the true extent of the battlefield. Instead Woodgate assumed that the terrain would be sloping away to the north of where he started to entrench and that the Boer attack would have to take place in full sight of the trenches. When the fog started to lift at about 07h00, Woodgate was horrified to find out that his position was between 70 and 180 meter from a crest line that dropped steeply away in the direction of the Boer positions. From his trenches he could not see any Boer attack before they reach the crest line. Although he tried to rectify this error by ordering some of his troops to dig in on the crest line, this was too little, too late. From the Boer perspective the counterattack was still executed under murderous fire from Spioenkop, but had the main trenches been dug in the correct place, it is questionable whether the Boer counterattack up Spioenkop would have succeeded. In this regard it is illuminating to read the eyewitness account of Deneys Reitz. After describing his climb up Spioenkop in the face of the British bullets, he contended that "it was marvelous that the Boers had got even thus far, for they had swarmed up the bare hillside in the face of a devastating fire, and they had pushed home the attack with such vigour that the narrow belt of rocks was thickly strewn with their dead" (Emslie 2009, p. 45). Had the British main trenches been on the crest line, this attack would have been made with much higher casualties. This might have led to a Boer retreat earlier in the day, handing victory to the British. More importantly, the alignment of Woodgate's main trenches allowed the Boers on Aloe Hill, to the east of his position, to fire directly into his trenches (Breytenbach 1973; Fig. 4.3). This proved to be a disastrous situation, and British soldiers were killed in scores, without any way in which to defend themselves, or to fight back effectively.

The Rains and Dark Night that Sapped the Energy of the Attackers

To further aggravate the British predicament, the intermittent rain that fell during the night (Amery 1902; Calitz and Pretorius 2011), and the pitch-black night, made the march up Spioenkop an extremely tiresome affair. Some soldiers dozed off the moment a halt was called. This situation, coupled to the steep slopes of Spioenkop against which the attacking force had to climb up to reach their objective, ultimately caused the soldiers to leave behind some of the heavy equipment used for entrenchment (Breytenbach 1973; Pakenham 1979).

When the British started to dig their trenches, they were exhausted from the murderous climb up Spioenkop, without proper equipment to dig the trenches, and to make matters worse, after only a couple of inches, they struck solid rock. Although the engineers had some pickaxes, these were soon blunt and useless. Under these conditions the trenches were dug only about 40 cm deep and improved by packing rocks and soil in front of the trenches (Ransford 1969; Breytenbach 1973).

The Inability of the British Artillery to Engage the Boer Guns

While the Boer gunners pounded the massed British soldiers in the main trench, the British gunners were unable to counter the Boer bombardment. Dye (2012) used modern mapping techniques, such as viewshed analysis, to show that the British gunners could not see or effectively engage the Boer guns from their position below

Spioenkop. The failure of the British to bring artillery with to the summit of Spioenkop, coupled to the inability of the British artillery to engage the Boer artillery from their position below Spioenkop, left the British soldiers exposed to the deadly fire of the Boer artillery on Aloe Knoll. This wreaked havoc among the tightly massed soldiers in the shallow trenches on the summit.

By nightfall, Woodgate was dead, and the British evacuated Spioenkop, handing the hill back to the Boers. Although it is difficult to accurately determine the exact amount of casualties, the British lost more than 1700 soldiers, while Boer losses amounted to about 200. The long list of British casualties was to no avail.

Conclusions

Although the battle of Spioenkop was not won or lost because of the physical geography of the battlefield, the outcome of this battle was certainly influenced by the physical geography of the area. What caused this battle to be a disaster and not a success, from a British perspective, was the inadequate and incorrect appreciation of the physical geography of the battlefield. It would be unfair to apportion blame for this only to the British generals who took and held Spioenkop. Inadequate mapping and reconnaissance, wrong positioning of the trenches because of the fog, the inability to dig in due to the heavy digging equipment they left behind during the wet and dark accent of Spioenkop, the shallow soil and hardness of the rocks, and the inability to engage the Boer guns with their own artillery due to the dead ground in which they were concealed turned opportunity into disaster. However, Woodgate could have rectified the lack of intelligence about the true nature of the battlefield by sending out scouts and changing his entrenchments accordingly. The price he had to pay for this was the terrible revenge enacted upon him by the physical geography of Spioenkop.

Therein lies the lesson to military commanders: ignore the importance of physical geography and be willing to pay the price in blood. That is a simple fact that holds true today, as it held true on 24 January 1900, on the summit of Spioenkop.

References

Ackermann, P. C. M. (1957). *Aardrykskundige invloede in die stryd tussen Brittanje en die Boere republieke met besondere verwysing na die Tweede Vryheidsoorlog.* Doktorale proefskrif, Universiteit van Suid Afrika.

Amery, L. S. (1902). *The times history of the war in South Africa, 1899–1902* (Vol. II). London: Sampson Low, Marston and Company, Ltd..

Breytenbach, J. H. (1973). *Die geskiedenis van die Tweede Vryheidsoorlog in Suid–Afrika, 1899–1902. Deel 3: Die stryd in Natal, Januarie-Februarie 1900.* Pretoria: Die Staatsdrukker.

Byers, E., & Guth, P. (2009). The battle of Attu: Mountain warfare at sea level. In H. Häusler & R. Mang (Eds.), *International handbook military geography* (Vol. 2). Vienna: Arbeidsgemeinschaft Truppendienst, Ministry of Defence and Sports.

Calitz, G., & Pretorius, F. (2011). Denys Reitz se deelname aan die Anglo Boereoorlog. The Anglo-Boer War 1899–1902. *A special edition of the Journal for Contemporary History, 36*(2), 90–111.

Doyle, P., & Bennett, M. R. (1999). Military geography: The influence of terrain in the outcome of the Gallipoli Campaign, 1915. *Geographical Journal, 165*(1), 12–36.

Doyle, P., & Bennett, M. R. (Eds.). (2010). *Fields of battle: Terrain in military history*. Dordrecht/ Boston/London: Kluwer.

Dye, M. (2012). *Spioenkop: A case study of the potential of modern mapping techniques to enhance understanding of historical battles*. Paper delivered at the 2012 GISSA Conference, Johannesburg, South Africa.

Emslie, T. S. (2009). *Adrift on the open Veld, the Anglo-Boer War and its Aftermath, 1899–1943*, by Denys Reitz. Cape Town: Stormberg Publishers.

Farwell, B. (1977). *The Great Boer war*. Hertfordshire: Wordsworth.

Galgano, F. A. (2011). Streams and military landscape. In F. A. Galgano & E. J. Palka (Eds.), *Modern military geography*. New York/London: Routledge.

Gnaser, G. (2009). Warfare in mountainous regions: The last days of the second world war in the Semmering area. In H. Häusler & R. Mang (Eds.), *International handbook military geography* (Vol. Volume 2). Vienna: Arbeidsgemeinschaft Truppendienst, Ministry of Defence and Sports.

Gray, C. S. (1999). Inescapable geography. Special issue on geopolitics, geography and strategy. (CS Gray & G Sloan, eds). *The journal of Strategic Studies, 22*(2/3), 161–177.

Jacobs, A., & Smit, H. A. P. (2004). Topographic mapping support in the South African military during the 20th century. *Scientia Militaria: South African Journal of Military Studies, 32*(1), 32–50.

Kestell, J.D. (1999). *Met die Boerekommando's* (trans: Botes, D.P.M.). Menlo Park: Protea Boekhuis.

Mang, R. (2009). Terrain evaluation of the Hannibalic Alpine Invasion Route. In H. Häusler & R. Mang (Eds.), *International handbook military geography* (Vol. 2). Vienna: Arbeidsgemeinschaft Truppendienst, Ministry of Defence and Sports.

Müller, A. U. (2006). Geographic key factors in global operations. In R. Mang & H. Häusler (Eds.), *International handbook military geography*. Vienna: Arbeidsgemeinschaft Truppendienst, Ministry of Defence.

Opperman, A. J. P. (1981). *Die slag van Majuba*. Johannesburg: Perskor Uitgewery.

Pakenham, T. (1979). *The Boer war*. London: Abacus.

Palka, E. J. (2011a). Military geography in the US: History, scope, and recent developments. In F. A. Galgano & E. J. Palka (Eds.), *Modern military geography*. New York/London: Routledge.

Palka, E. J. (2011b). Afghanistan: Operation enduring freedom and military geographic challenges. In F. A. Galgano & E. J. Palka (Eds.), *Modern military geography*. New York and London: Routledge.

Palka, E. J. (2011c). Iraq and operation Iraqi freedom: A military geography. In F. A. Galgano & E. J. Palka (Eds.), *Modern military geography*. New York/London: Routledge.

Perry, J. M. (1996). *Arrogant armies: Great military disasters and the generals behind them*. New York: Johan Wiley & Sons.

Pretorius, F. (2002). *Die invloed van landskap op die Anglo Boere Oorlog*. Unpublished lecture.

Puckett, R. M. (1994). The Rapido River crossings. *Military Review LXXIV, 1*, 73–76.

Ransford, O. (1969). *The Battle of Spion Kop*. London: John Murray.

Scholtz, G. D. (2000). *The Anglo-Boer war, 1899–1902*. Pretoria: Protea Book House.

Scholtz, L. (2005). *Why the Boers lost the war*. Houndmills: Palgrave Macmillan.

Sleigh, D. (1979). *Ruiters teen die Ryk: Die verhaal van die Tweede Vryheidsoorlog*. Goodwood: Nassou Beperk.

Smit, H. A. P. (2011). Methuen's Northern Cape Campaign, Anglo-Boer War. In F. A. Galgano & E. J. Palka (Eds.), *Modern military geography* (pp. 1899–1902). New York/London: Routledge.

Smurthwaite, D. (1999). *Hamlyn history: The Boer war, 1899–1902*. London: Hamlyn.

Tate, J. (2006). Terrain analysis for decision making. In R. Mang & H. Häusler (Eds.), *International handbook military geography*. Vienna: Arbeidsgemeinschaft Truppendienst, Ministry of Defence and Sports.

Van der Waag, I. (2013). The battle of Sandfontein, 26 September, 1914: South African military reform and the German South-West Africa campaign, 1914–1915. *First World War Studies 4*(2), pp. 1–25.

Wessels, A. (1991). *Die Anglo-Boereoorlog, 1899–1902: 'n Oorsig van die verloop van die stryd*. Bloemfontein: Oorlogsmuseum.

Winters, H. A., Galloway, G. E., Jr., Reynolds, W. J., & Rhyne, D. W. (1998). *Battling the elements: Weather and terrain in the conduct of war*. London: The Johns Hopkins University Press.

Chapter 5
Military Geoscientific Materials for Excursions to Theatres of First World War in France and Belgium

Reinhard Mang and Hermann Häusler

Introduction

This paper presents military geoscientific aspects of excursions to the main theatres of First World War in Western Europe – the Verdun area in France and the Flanders area in Belgium. The first part deals with the geospatial preconditions which led to the so-called trench war, and the second part will give some geological details of these battlefields. Figure 5.1 provides an overview of the geographic and political situation in Central Europe at the beginning of the WWI in 1914.

In the west, the United Kingdom and France (blue) opposed the so-called Central Powers of Germany and Austria-Hungary (red). These two opposing parties were separated by a corridor of neutral states, the Netherlands, Belgium and Luxemburg in the north and partly by Switzerland in the south.

This boundary situation of the Central European countries dates back to the war between Germany and France in 1870–1871. After winning that war, Germany had annexed Alsace and the German-speaking part of Lorraine. As a consequence, France considered how to avoid another German invasion in the future, and Raymond Serré de Rivière, general of the engineers, proposed a line of strongholds paralleling the new border. This line was later on called the "Iron Line", and four outstanding cities in France were representative for it, namely, Belfort, Epinal, Toul and Verdun (Fig. 5.2).

R. Mang
Institute for Military Geography of the Austrian Armed Forces, Vienna, Austria

H. Häusler (✉)
University of Vienna, Department of Environmental Geosciences, Vienna, Austria
e-mail: hermann.haeusler@univie.ac.at

P. L. Guth (ed.), *Military Geoscience*, Advances in Military Geosciences,
https://doi.org/10.1007/978-3-030-32173-4_5

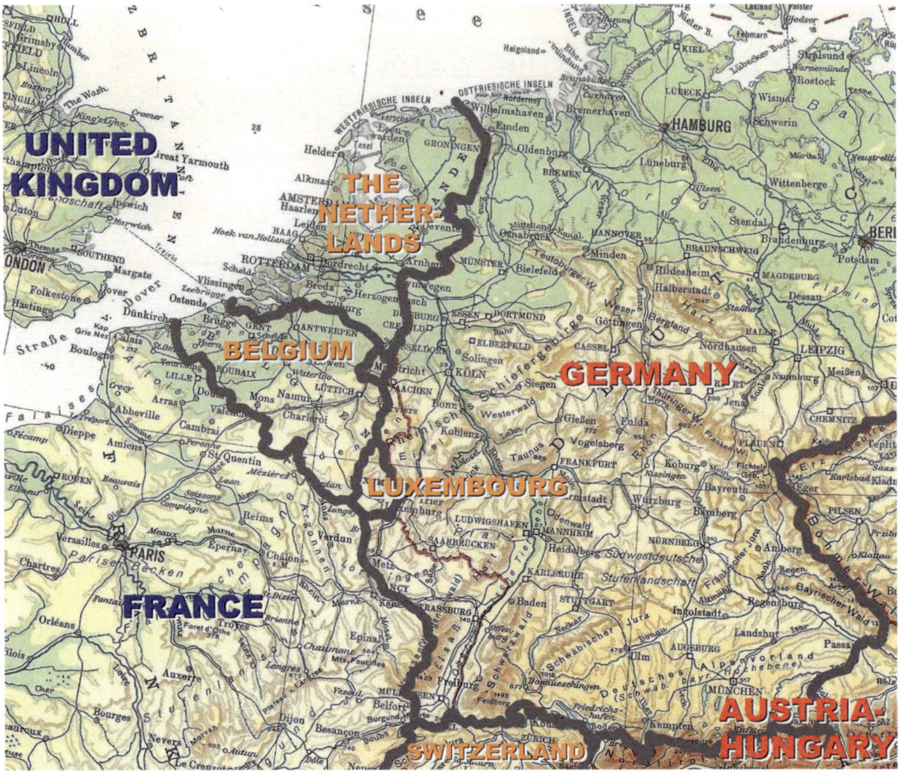

Fig. 5.1 The opposing countries (red and blue) of Central Europe in 1914. (Base map from historical atlas published around 1900)

From Mobile Warfare to the "Trench War"

The German general staff was of the opinion that there was only one way to successfully defeat France again – invading France by marching through the neutral countries of Luxemburg and Belgium. A very strong "right wing" of the German armies would invade Belgium and Luxemburg and then enter France, surrounding Paris in the west and then appearing in the back of the "Iron Line". Germany's plan was to defeat France within a few weeks and only then bring considerable military forces to the eastern front against Russia to avoid the dreaded "two-front war". This plan was later on called the "Schlieffen Plan", termed after the former German chief of staff Field Marshal Alfred von Schlieffen (1833–1913), who always stressed "Keep the right wing strong".

Figure 5.3 displays nine important steps of the Schlieffen Plan which were presented at the ICMG Conference in Quebec in 2007. On August 2, 1914, the Germans started an offensive in the north, whereas the French twice attempted to invade

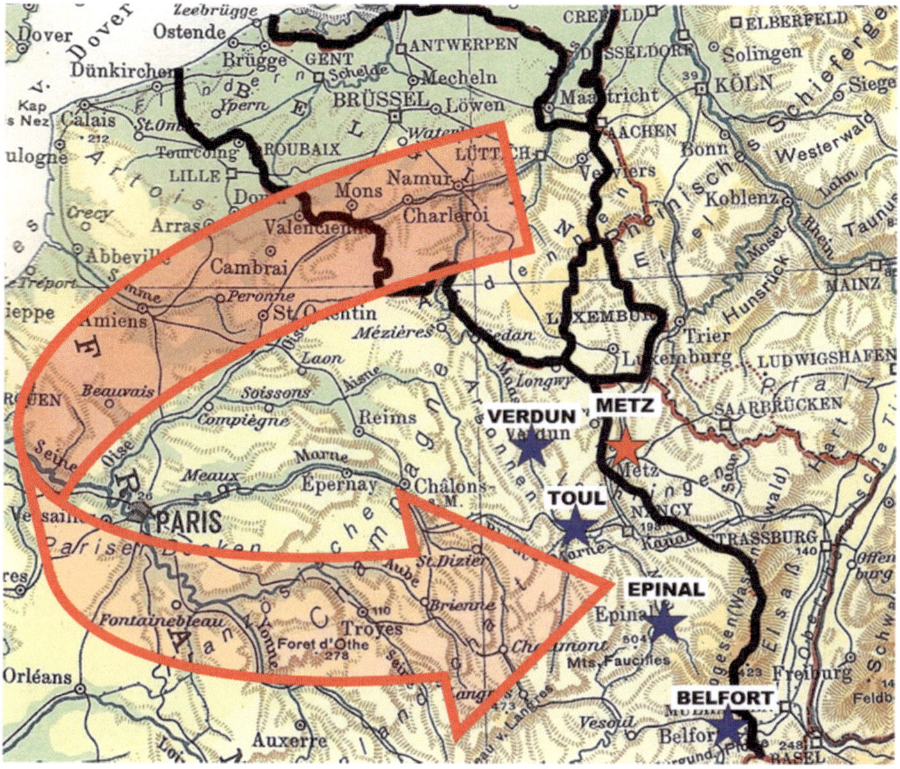

Fig. 5.2 The "Iron Line" of French fortresses (blue stars) and the German "Schlieffen Plan" (red arrow) in 1914. (Base map from historical atlas published around 1900)

Alsace in the south. Later on, the Belgian army retreated into the fortress of Antwerp, and therefore the Germans diverted two corps from their right wing to besiege Antwerp.

When the Russian army invaded Eastern Prussia, further, two corps were removed from the right wing to support the German armies in Eastern Europe. In addition, yet another German corps was removed for the besieging of Maubeuge on August 24. At this point, the German right wing had been weakened by a total of five corps, which is the equivalent of about one whole army, a possibility which had not been taken into consideration. It is likely that at this stage the Germans recognized that they did not have sufficient strength to surround Paris from the west as planned. As they moved southwards to pass Paris to the east, they opened a flank to the still effective French and British forces. The southernmost line the Germans achieved was south of the river Marne. Due to German communication problems, contradictory assessments of the situation and counterattacks by the Allied forces, the German forces were forced back, and it became evident that the "Schlieffen Plan" had failed. This withdrawal is known as "The Marne Miracle".

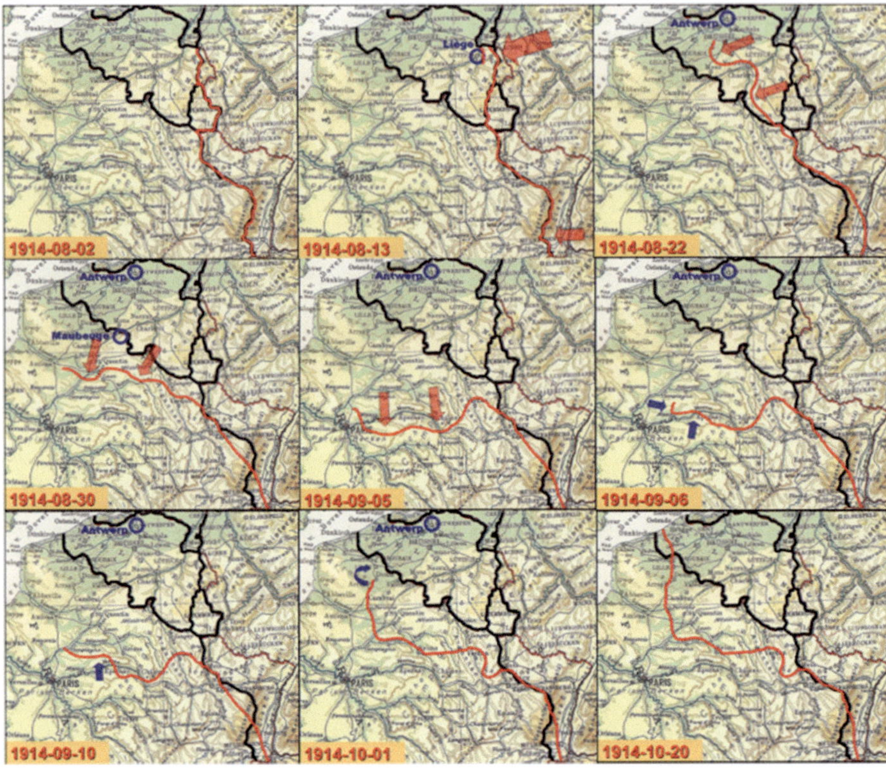

Fig. 5.3 Important steps of German operations on the Western Front dating from August 2 to October 20, 1914. The red line indicates the German front. Red arrows show attack, and blue arrows show the retreat of the German armies. (Base map from historical atlas published around 1900)

Later on a number of attacks and counterattacks occurred on the German northwestern flank. The opposing parties tried to outflank each other in the direction of the North Sea, and later on these operations were called "The Race to the Sea".

The Race to the Sea finally stopped at the harbour of Nieuwpoort, when the Belgians opened the locks separating the River Yser from the North Sea. By flooding the flat area between Nieuwpoort and Ypres (Fig. 5.4), they stopped the German troops, and for the next 4 years, Flanders and Verdun became the centres of the so-called trench war (Fig. 5.5).

In winter 1914, the Western Front stabilized along a line bordering Northeast France (Belfort–St. Die–Nancy–Toul–Verdun–Reims–Soissons–Arras, up to Lille and Ypres; see Fig. 5.6), and the armies tried to use the properties of the ground to their best advantage. Material for military geographic and geologic excursions along the Maas Valley and to Verdun in France and along the Yser valley and the Ypres Salient in Belgium is described using evidence acquired from military atlases, military historic excursion guides 1:140,000, historic military geology maps

Fig. 5.4 Map of area situated below sea level (blue) and troops in Northern Belgium in October 1914 (modified after Birken and Gerlach 2002, left; courtesy of Philathek publishers). Harbour of Nieuwpoort with algae-covered concrete ramp indicating height of tidal change (right; photo taken by Reinhard Mang, 1 July 2006)

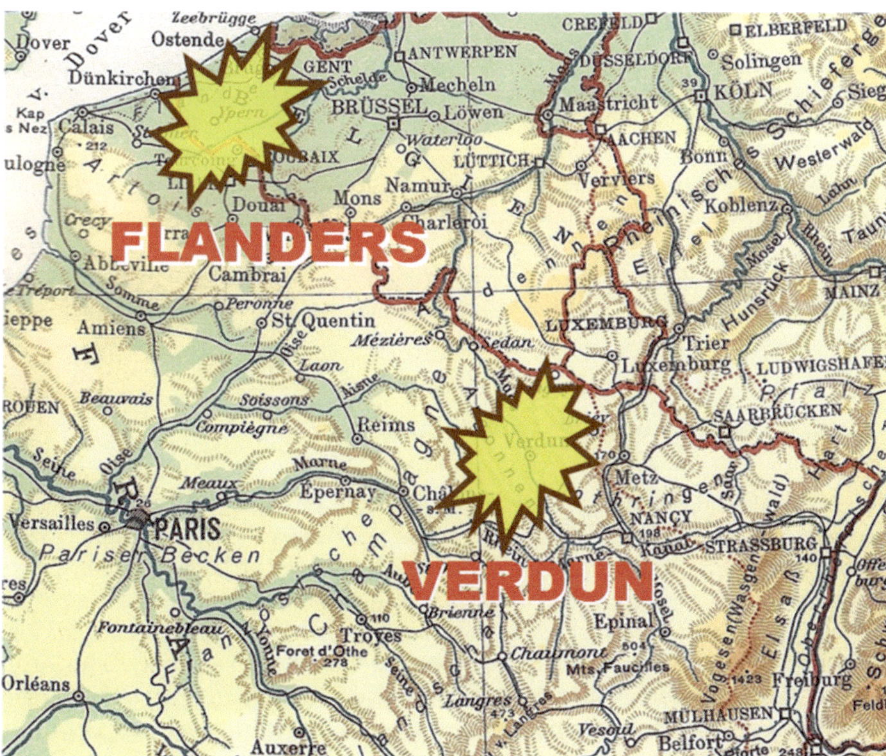

Fig. 5.5 From October 20, 1914, Flanders and Verdun became the centres of trench warfare in Western Europe. (Base map from historical atlas published around 1900)

Fig. 5.6 Overview of deployment of military geology groups to the "Armeeabteilung Gaede", "Armeeabteilung Falkenhausen", "Armeeabteilung Strantz" and the "Herzog Albrecht's Fourth Army". (Modified after Birken and Gerlach (2002, courtesy of Philathek publishers))

1:300,000, local terrain evaluation by military geologists of the First World War, (Kraus 1918; Wilser 1927), exhibitions of war museums and excursion guides to the battlefields. For the study of military history, we refer to the information provided by war museums and to modern atlases (Birken and Gerlach 2002; Klauer 2004; both written in German).

Military Geology of the "Trench War"

General field guides to the geology of the Western Front have been published in German (Verdun: Sturm 1923; Alsace: Kraus and Wagner 1924; Lorraine: Kraus and Klüpfel 1925) and in English (Doyle 1998, 2000; Doyle et al. 2000, 2002; Chasseaud 2002). German war geologists published their experiences in a paper on "War Geology" which was edited by the chief of war cartography in January 1918 (Chef des Kriegsvermessungswesens 1918) and another by Dr. Ernst Kraus, who headed a military geology unit from 1916 on (Kraus 1918, 1919). In addition, the war geologist Dr. Walter Kranz reported case studies of mining and trenching (Kranz 1935, 1936a, b, 1937), and he also coedited a textbook on military geology (Bülow et al. 1938; see also G. Keller 1936).

The organisation of the British military geology at the Western Front in May 1916 comprised two geologists of the Royal Engineers. Lieutenant (later Captain) William Bernhard Robinson King was working as a staff officer to serve with the chief engineer (later retitled engineer-in-chief) of the British Expeditionary Force. Major (later Lieutenant Colonel) Tannatt William Edgeworth David provided specialized geotechnical maps illustrating the relative suitability of the ground for dugout construction. In 1915 9 British tunnelling companies and in 1916 25 Allied tunnelling companies comprised in total 25,000 men who were actively engaged in mining (Rose et al. 2000). In late 1916 and in 1918, three more geologists were attached to tunnelling companies (Doyle et al. 2000).

The organisation of the German military geology at the Western Front comprised about 60 war geologists organized into geology groups. These military geology groups were attached to the geodetic survey ("Kriegsvermessungswesen", Häusler 2000), and therefore each geology group advised an army or army corps in the west. In total, several thousand military geological expert opinions, both oral and written, explained soil and subsoil conditions for water supply, drainage of trenches, aggregates, dugouts and mining from 1915 to 1918.

From December 1916 onwards, one geology group was attached to the General Gaede's Armeeabteilung (later termed "Armeeabteilung B" or short A.A.B.; Fig. 5.6). The group was located at Colmar and comprised five military geology units at the front. At the Western Front, 12 German military geologists, 12 additional geologists, 9 technical personnel and 25 privates provided 13 military geology maps 1:10,000, 11 military geology maps 1:25.000 and about 1000 military geology expertises (Kraus 1919).

Another military geology group was attached to General Falkenhausen's Armeeabteilung (later "Armeeabteilung A" or A.A.A.) and was headed by Major Dr. Walter Kranz and Lieutenant Dr. W. Wagner.

Verdun

Although someone visiting Verdun will be presented with renovated trenches and a large selection of books, brochures, maps and videos, there is no specific information on the terrain itself. This leads to a lack of understanding of the military geography and for the military geology of the battlefield between the Argonne Woods and Woëvre plain, where 360.000 French and 330.000 German troops died fighting for only 5 kilometres of terrain between February 1916 and August 1917. Geologically, this area is composed of limestone and marlstone of Upper Jurassic age. Due to tectonic tilting of these formations, limestone beds more resistive to erosion form steeper flanks bordering the Maas Valley in the west, which are termed "Côtes de Meuse" (Fig. 5.7).

In November 1915, one military geology group was attached to General von Strantz's Armeeabteilung (later "Armeeabteilung C"; in short A.A.C.) in the

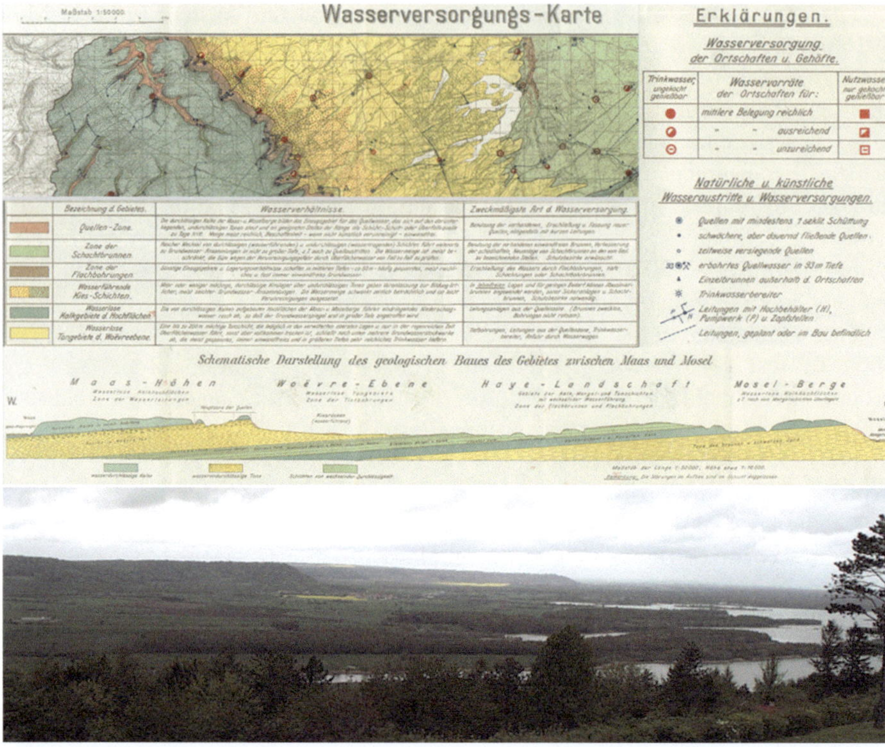

Fig. 5.7 Section of the geological map at original scale 1:50.000 with legend and explanations for drinking water supply ("Wasserversorgungs-Karte", Chef des Kriegsvermessungswesens 1918). The geological profile depicts escarpments of permeable, karstified limestone of Jurassic age overlying impermeable claystone forming gentle hills. View on "Cótes de Meuse" in the Maas Valley. (Photo taken by Hermann Häusler, 5 May 2004)

Verdun–Metz sector of the German front. It was headed by university professor Lieutenant Dr. Hans Philipp and comprised 6 military geology units of 28 military geologists in total. For tunnelling, mining and water supply, about one dozen German war geologists assisted General Stranz's army in the Verdun sector. German mining and galleries of the front sector of General Gallwitz ("Gallwitz-Tunnel") are presented on numerous panels in the field.

Flanders

In November 1915, Dr. Wilfried von Seidlitz headed the military geology group attached to Duke Albrecht's Fourth Army. This military geology group reported to the staff officer of the geodetic survey of the Fourth Army high command. The

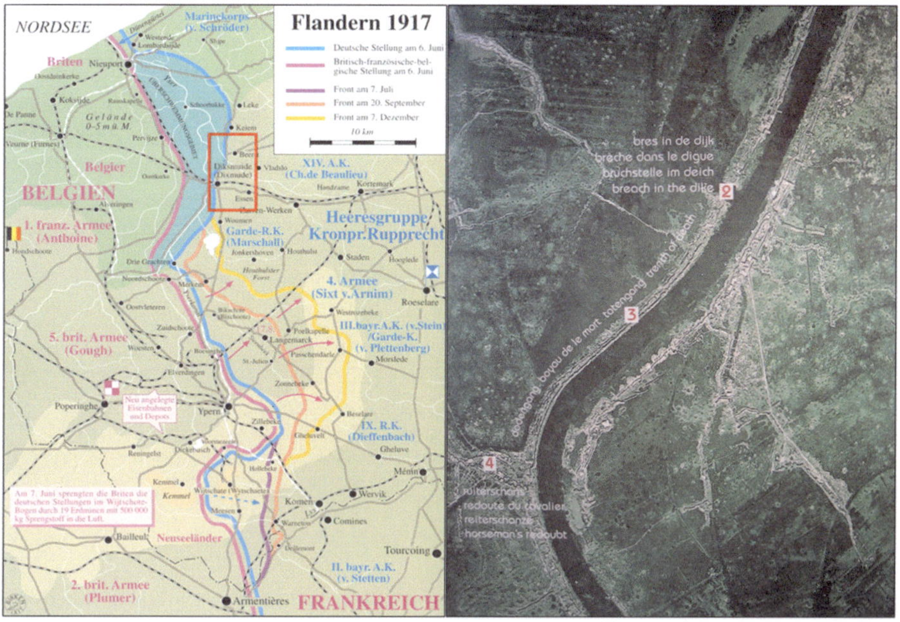

Fig. 5.8 Inundation area of the River Yser south of Nieuwpoort harbour (modified after Birken and Gerlach 2002, left; courtesy of Philathek publishers). Trenches located at both sides of the River Yser north of Diksmuide (right; panel exposed at Dodengang museum with "trench of death"; photo taken by Reinhard Mang, 5 July 2006)

group was composed of six military geology units which provided both expertises on flooding and a military geology atlas of Flanders (Fig. 5.8).

Detailed geologic maps are available as reprints at the geological survey in Brussels. Formations in Flanders are termed after sites situated in Belgium (e.g., Yper and Mont Panisel, near Mons). Consisting of mostly sand and clay, the formations were a large factor in the success of trenching for simple geologic reasons: clay is more or less impermeable, stopping the flow of water. Thus, overlying permeable sands act as an aquifer, and therefore – basically – trenches in the clay were wet, whereas trenches in the sand were dry (Fig. 5.9).

The valleys south of Ypres incise the Tertiary and Quaternary formations about 50 metres above sea level. This geologic situation was quite complicated because the Ypres Formation and the overlying Paniselian Formation comprised several aquifers, which considerably complicated mining activities (Fig. 5.10). From March 1916 the British army started a mining attack along the Messines–Wytschaete Ridge. Following the advice of their military geologists to dig trenches with adequate drainage, British miners started from the valley floors horizontally in the dry Ypres Clay and covered distances of more than 5400 m. Doyle (1998) and Doyle

Fig. 5.9 Section of the Belgian geological map at original scale 1:40,000, sheet Neuve–Église–Messines (Institut cartographique militaire 1900), with legend of marine formations of Lower Eocene age. Red rectangles mark the important localities Wytschaete and Hollebeke south of Ypres

et al. (2000) presented examples of good and poor trenches and dugouts on the Western Front in relation to the underlying geology. Good positions would be well drained and dry, whereas poor positions would be liable to flooding through inadequate drainage or water seepage.

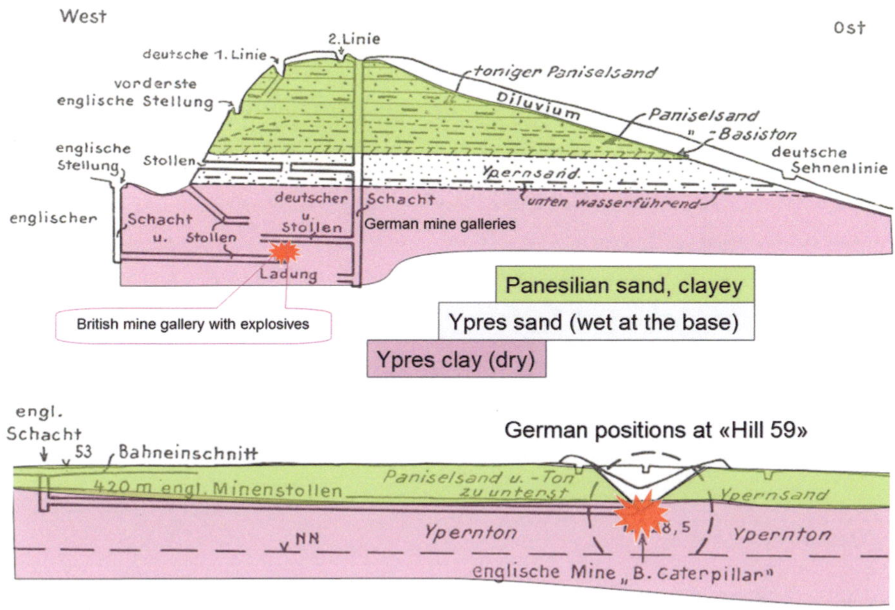

Fig. 5.10 Schematic geologic sections across the Messines–Wytschaete Ridge in 1917 (above). The German mine galleries were cut into the sand and clayey sand of the Paniselian Formation, whereas the British mine galleries were cut into the underlying dry Ypres Clay. The "Caterpillar Crater" at Hill 59, northeast of Wytschaete, resulted from the explosion of 31 tons of explosives. (Modified after Bülow et al. 1938)

Along the Messines–Wytschaete Ridge, a total of 19 galleries were cut into the Ypres Clay, far below the German positions (Oldham 2003). Dr. Walter Kranz reported that the German troops at that time disregarded the advice of the German war geologists who tried to explain the advantages and disadvantages of the mining situation in detail (Bülow et al. 1938, p. 69; see Fig. 5.11, this paper). Finally, Wytschaete has become a synonym for the purposeful, simultaneous blasting of 500 tons of explosives in 19 mines of the Battle of Messines on June 7, 1917. Aerial reconnaissance flights clearly reveal the damage caused by artillery and gallery explosions. Lying in what is today a recreational park, the Caterpillar Crater is situated opposite Hill 60 at an altitude of 60 m above sea level (Fig. 5.12).

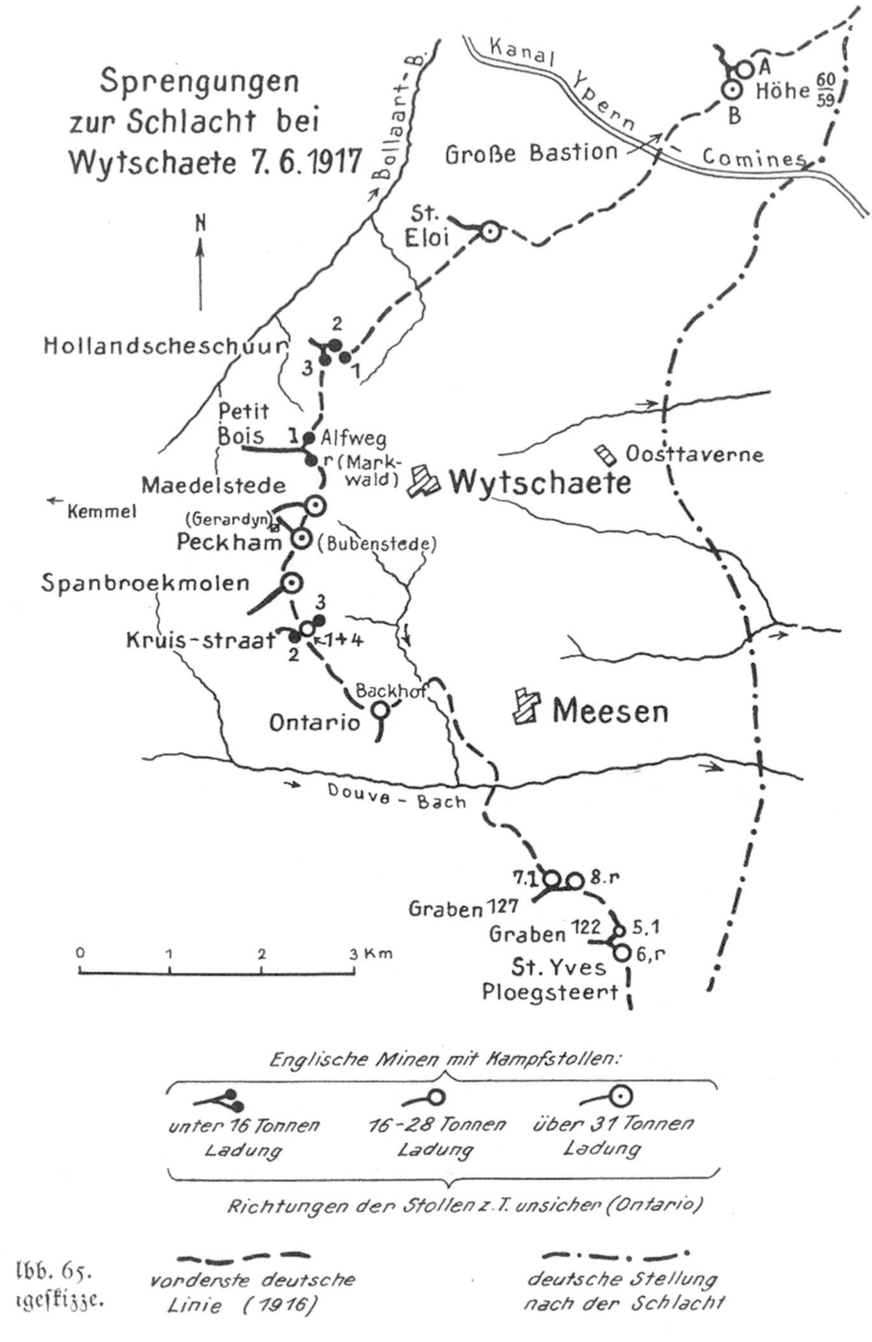

Fig. 5.11 Sketch of the German positions and mines blown up at the battle of Wytschaete on June 7, 1917. Legend for British mines and mine galleries ("Englische Minen mit Kampfstollen"): black dots, mines with less than 16 tons of explosives; white circle, mines with 16–28 tons; white circle with centre point, mines exceeding 31 tons of explosives. "Hill 59" and "Hill 60" are located northeast of Wytschaete (Bülow et al. 1938, see Cave 2004)

Fig. 5.12 Comparison between British trench maps and recent coloured orthophoto of the "Caterpillar Crater" near Hollebeke south of Ypres, presented at Flanders Fields Museum, Ypres (above), and present landscape with the "Caterpillar Crater lake". (Below; photos taken by Reinhard Mang, 30 June 2006)

Acknowledgements The authors are grateful to two anonymous reviewers for constructive commentary and suggestions for improving the manuscript.

References

Birken, A., & Gerlach, H. H. (2002). *Atlas und Lexikon zum Ersten Weltkrieg. I Karten* (p. 114). Hannover: Philathek-Verlag.

Cave, N. (2004). *Battleground Europe, Ypres: Hill 60* (p. 160). Barnsley, South Yorkshire: Leo Cooper.

Chasseaud, P. (2002). British, French and German mapping and survey on the Western front in the first world war. In P. Doyle & M. R. Bennett (Eds.), *Fields of battle – Terrain in military history, The GeoJournal Library 64* (pp. 171–204). Dordrecht: Kluwer Academic Publishers.

Chef des Kriegs-Vermessungs-Wesens. (Hrsg.). (1918). *Kriegsgeologie*. Druckerei des General-Gouvernements in Belgien, Brüssel, p. 78.

Doyle, P. (1998). *Geology of the Western front, 1914–1918. Geologists´ Association Guide 61* (p. 79). London: The Geologists´ Association.

Doyle, P. (2000). A field guide to the geology of the British sector of the Western Front, 1914–18. In E. P. F. Rose & C. P. Nathanail (Eds.), *Geology and warfare: Examples of the influence of terrain and geologists on military operations* (pp. 381–412). London: The Geological Society of London.

Doyle, P., Matthew, R. B., & Cocks, F. M. (2000). Geology and warfare on the British sector of the Western Front 1914–18. In E. P. F. Rose & C. P. Nathanail (Eds.), *Geology and warfare: Examples of the influence of terrain and geologists on military operations* (pp. 179–235). London: The Geological Society of London.

Doyle, P., Bennett, M. R., Macleod, R., & Mackay, L. (2002). Terrain and the Messines Ridge, Belgium, 1914–1818. In P. Doyle & M. R. Bennett (Eds.), *Fields of battle – Terrain in military history, The GeoJournal Library 64* (pp. 205–224). Dordrecht: Kluwer Academic Publishers.

Häusler, H. (2000). *Die Österreichische und Deutsche Kriegsgeologie 1914–1918. Informationen des Militärischen Geo-Dienstes, 75* (p. 161). Wien: Institut für Militärisches Geowesen.

Institut cartographique militaire. (1900). Carte géologique de la Belgique, 1:40.000, no. 95, Neuve-Église-Messines (reprint), Brussels.

Keller, G. (1936). Die geologischen Vorraussetzungen für den Minenkrieg im Wytschaetebogen. *Zentralbl Min Geol Pal, 1936B*, 235–242.

Klauer, M. (2004). *Militärgeschichtlicher Reiseführer zu den Schlachtfeldern des Ersten Weltkrieges in Flandern und Nordfrankreich* (p. 191). Velbert: Gesellschaft für Druck und Veredelung mbH.

Kranz, W. (1935). Minierkampf und Kriegsgeologie im Wytschaetebogen. *Vierteljahreshefte für Pioniere, 2*, 166–181.

Kranz, W. (1936a). Allgemeine Lehren aus den Minierkämpfen im Weltkrieg: Die Wytschaete-Sprengungen bei Hollandscheschuur und Petit Bois; taktische Wirkungen der Sprengungen am 7. Juni 1917. *Sonderbeilage zu "Deutsche Wehr", 40*(31), 1–6.

Kranz, W. (1936b). Kämpfe im Wytschaetebogen und Kriegsgeologie. *Nachrichten aus Württemberg, 209*.

Kranz, W. (1937). Der Kampf um Wytschaete. *Deutsche Pionier-Zeitung, 14*, 13.

Kraus, E. (1918). Geologische und kriegsgeologische Übersicht über das Gebiet der Armeeabteilung B. *Vermessungsabteilung, 13*, 17.

Kraus, E. (Ed.). (1919). *Die Kriegsgeologie des Oberelsaß* (p. 176). Vermessungsabteilung 13: Berlin.

Kraus, E., & Klüpfel, W. (1925). Lothringen. In J. Wilser (Ed.), *Die Kriegsschauplätze 1914–1918 geologisch dargestellt* (Vol. 2, p. 212). Berlin: Borntraeger.

Kraus, E., & Wagner, W. (1924). Elsass. In J. Wilser (Ed.), *Die Kriegsschauplätze 1914–1918 geologisch dargestellt* (Vol. 1, p. 154). Berlin: Borntraeger.
Lais, R. (1925). Zwischen Maas und Mosel. In J. Wilser (Ed.), *Die Kriegsschauplätze 1914–1918 geologisch dargestellt* (Vol. 3, p. 116). Berlin: Borntraeger.
Oldham, P. (2003). *Battleground Europe, Ypres: Messines Ridge; Messines – Wytschaete – St. Eloi* (p. 144). Barnsley, South Yorkshire: Leo Cooper.
Rose, E. P. F., Häusler, H., & Willig, D. (2000). Comparison of British and German military applications of geology in world war. In E. P. F. Rose & C. P. Nathanail (Eds.), *Geology and warfare: Examples of the influence of terrain and geologists on military operations* (pp. 107–140). London: Geological Society.
Sturm, F. (1923). Vor Verdun. In J. Wilser (Ed.), *Die Kriegsschauplätze 1914–1918 geologisch dargestellt* (Vol. 4, p. 44). Berlin: Borntraeger.
von Bülow, K., Kranz, W., Sonne, E., Burre, O., & Dienemann, W. (1938). *Wehrgeologie* (p. 164). Leipzig: Quelle & Meyer.
Wilser, J. (1927). Die Kriegsschauplätze 1914–18 geologisch dargestellt. *Forschungen und Fortschritte, 3*, 77.

Reinhard Mang

1950: Born in Vienna, Austria
1971: Studies of Geography/Cartography at the University of Vienna, Austria
1979: Federal Ministry of Defence, Operations Department, Military Geography
1992: Head of Military Geographic Service
1992: Postgraduate studies (GIS) at the Technical University of Vienna
1997: Head of the Institut für Militärisches Geowesen (Institute for Military Geography) of the Austrian Armed Forces, Vienna
2013: Retired
2014: Deceased

Hermann Häusler

1953: Born in Linz, Austria
1972: Studies of Geology, Engineering Geology and Soil Science
1978: Dr. phil. (Geology, University of Vienna), Associate Professor at the Institute of Geology, University of Vienna; Postgraduate training at a Technical Bureau of Applied Geology and at a Technical Bureau of Hydrogeology
1987: Postgraduate studies (Photogrammetry) at the Technical University of Vienna
1989: Professorship at the University of Vienna
1998: Colonel of advanced military technical service at the Institut für Militärisches Geowesen (Institute for Military Geography) of the Austrian Armed Forces, Vienna
2005: Militia expert at Institut für Militärisches Geowesen (Institute for Military Geography) of the Austrian Armed Forces, Vienna
2018: Brigade General (Militia), retired

Chapter 6
The Cadore Offensive: Theoretical Military Geography Considerations

Mauricio Nicolas Vergara and Aldino Bondesan

Introduction

Italy declared war against Austria-Hungary on 23 May 1915 and joined the Allied forces about 1 year after the assassination of Archduke Franz Ferdinand of Austria in Sarajevo. The theater of war between the Italian Kingdom and the Austro-Hungarian (Hapsburg) Empire extended mainly through the Eastern Alps, from the Swiss border to the Adriatic Sea (Fig. 6.1). By the end of the war, on this front, Italian casualties numbered 651,000 deaths and 953,000 wounded, and Austria-Hungary numbered 404,000 deaths and 1,207,000 wounded.

Due to Italy's military alliance, as well as political, economic, and irredentist national interests, the Italian military was forced to take an offensive stance. The operational design of General Cadorna, who was the chief of staff of the Royal Italian Army, identified the country's key strategic objectives beyond the Julian Alps and the "Carso" (the "Classic Karst" area along the Italian border with modern-day Slovenia; Fig. 6.1). The bulk of Italian forces was therefore deployed in the directions of those aims. This operational design determined General Cadorna's planning from the summer of 1914 until the autumn of 1917, when Austro-Hungarian forces swept into the plains of northern Italy.

Just before the Italian declaration of war, however, the Julian and Carso strategy was momentarily abandoned. This was mainly because, after a careful reassessment of the strategic and organizational situations of the two armies, General Cadorna realized that the Austro-Hungarians would have deployed faster than the Italian troops. In the document *Variations to the directives of 1st September 1914*, issued on 1 April 1915, General Cadorna thus established for the Royal Italian Army a general strategic defensive stance that was planned to last from the declaration of war until its full mobilization. The only exception to this strategic defensive stance

M. N. Vergara (✉) · A. Bondesan
Department of Historical, Geographical and Antiquity Sciences, University of Padova, Padova, Italy

P. L. Guth (ed.), *Military Geoscience*, Advances in Military Geosciences,
https://doi.org/10.1007/978-3-030-32173-4_6

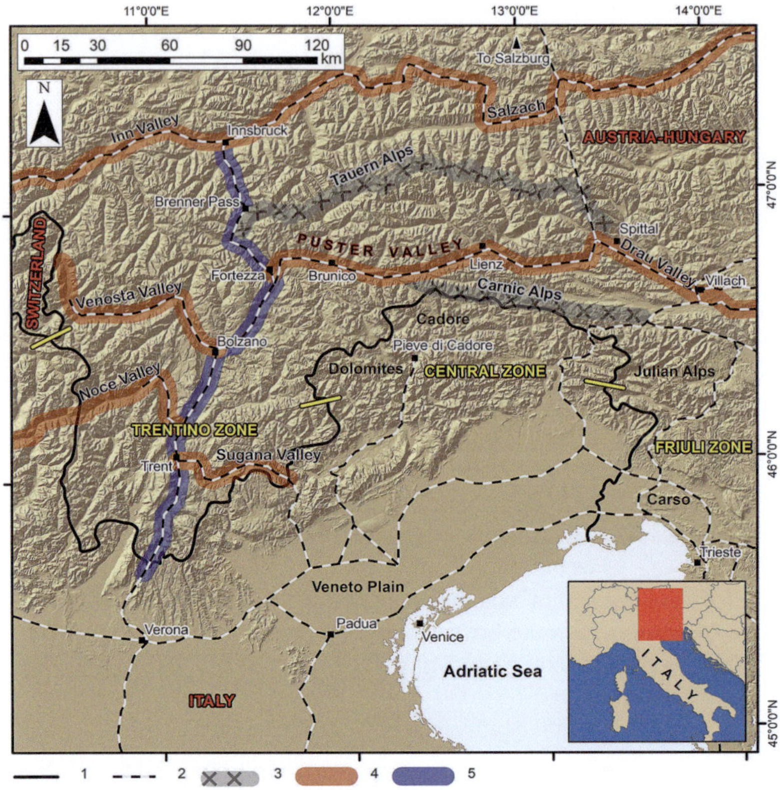

Fig. 6.1 Location of the Puster Valley inside the theater of war between the Austro-Hungarian Empire and the Italian Kingdom. Geographical features of strategic interest inside the Eastern Alps include (1) state border 1866 - 24 May 1915 (yellow bar: boundary of General Cadorna's areas of operation); (2) main railways before 24 May 1915; (3) topographic barrier; (4) lateral corridor; (5) main north-south corridor

concerned the 4th Army, which was deployed in the mountainous region of Cadore (Fig. 6.1). This Army was assigned the first objective of the Cadore offensive: to reach the Puster Valley. The 4th Army thus became the only unit that from the beginning of the war had an offensive task beyond the state border of strategic, and not just tactical, importance. More importantly, such an attack toward the Puster Valley was the first time in history that the Italian Army intended, from the start of operations, to cross the political border of Italy and to take possession of an important objective beyond it (Di Martino and Cappellano 2007).

The first step for the 4th Army's commander, General L. Nava, was to overcome the Austro-Hungarian "sbarramenti" that blocked the routes to the Puster Valley (Fig. 6.2). The term "sbarramenti" refers to the Austro-Hungarian blockade of mountain passes and valleys. Before the war the blockade consisted of forts, but as war with the Italian Kingdom became more likely, these forts were reinforced by numerous field fortifications.

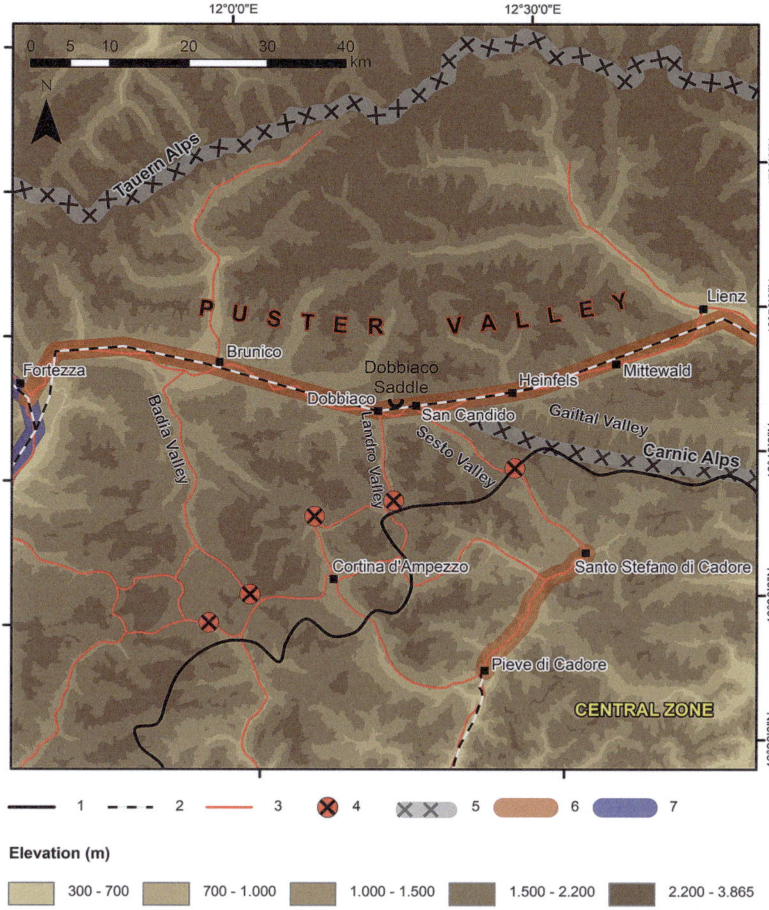

Fig. 6.2 Topographical map of the northern region of the Central Zone: (1) state border 1866 - 24 May 1915; (2) main railways before 24 May 1915; (3) main roads before 24 May 1915; (4) main Austro-Hungarian "sbarramenti" protecting the Puster Valley; (5) topographic barrier; (6) lateral corridor; (7) main north-south corridor

During the summer of 1915, the 4th Army launched its biggest offensive against the "sbarramenti" without any positive result. As supplies continued flowing to this front from both sides and Austro-Hungarian defenses grew stronger, also the battles grew bloodier. The Italians tried also to outflank the Austro-Hungarian "sbarramenti," and this made battles extend all along the border between the two countries, from the valleys to the mountainsides, on glaciers, and on the summit peaks, thus becoming one of the most important instances of mountain warfare in military history.

By the beginning of autumn 1915, General Cadorna realized that attritional warfare had been established even in the area assigned to the 4th Army. This dashed his hopes of reaching the Puster Valley, which would have provided a line of penetration into Austro-Hungarian territory. This induced him to redirect the 4th Army to a mainly defensive strategy, and offensive action was designated exclusively for the great mass of men concentrated in the direction of the Julian Alps and the Carso (Fig. 6.1). This situation prevailed until October 1917, when the Austro-Hungarians, no longer engaging the Russians, advanced with German support, from the Julian Alps onto the Veneto Plain. This forced the withdrawal of the majority of the Italian units, including the 4th Army, from Cadore.

The failure of the attack on the Puster Valley has occupied an important place in the debates over the Royal Italian Army's operations in what many considered the 4th Italian War of Independence. Discussions included, for example, General Cadorna's planning (e.g., his operation plan and the operational design); General Nava's long and excessively methodical preparation for attack; the size and growth of Austro-Hungarian forces occupying the border from early in the conflict; the Serbian-Russian attack that should have been launched at the same time Italy joined the war, but which did not happen; the 4th Army's lack of technical siege equipment; and the confusing nature of General Cadorna's orders and directives (see, e.g., Cadorna 1921 and 1950; Nava 1922; Romolotti 1978; Berti 1982; Pieri 1998; Di Martino and Cappellano 2007 for further discussion of these topics).

Although almost all analyses of the Italian operations agree on the immense obstacles the geography presented, not only for the Cadore offensive but for offensive actions along the entire Alpine front, there has yet to be any specific or systematic study of the subject. To contribute further to our understanding of the failure of the Cadore offensive, this paper focuses on the geography of the area, considering its military significance from a theoretical point of view.

Our study begins with the theoretical assessment of the military significance and influence of geography that General Cardona used as the basis for his operation plan. We direct our attention solely to the offensive aspects of his assessment, because the operation plan followed by the Royal Italian Army when joining the war defined an offensive strategy for Cadore.

After identifying General Cadorna's theoretical assessment (section "Cadorna's Theoretical Military Geography Assessment of the Central Zone of Operations (Cadore)") and describing the historical-geographical framework (section "Historical-Geographical Framework of the Area Relevant to the Cadore Offensive"), this paper contributes some theoretical military geography considerations based on the "Study Guide for Military Geography" of General C. Porro (1898) and on some early twentieth-century military studies (section "Theoretical Military Geography Considerations Relevant to the Cadore Offensive"). In section "Results", General Cadorna's theoretical assessment is examined from the perspective of theoretical military geography considerations and certain logistical considerations about the Cadore offensive voiced by some senior Italian officers who served during WWI (section "Logistical Considerations of the Cadore Offensive").

Cadorna's Theoretical Military Geography Assessment of the Central Zone of Operations (Cadore)

The operation plan which shaped the Royal Italian Army participation in WWI was the sole creation of General Cadorna (Cadorna 1950). Any military operation plan is developed prior to the initiation of war and must, therefore, be adapted to any contingencies that occur in the time between its development and the actual initiation of war. A military geography assessment must be at the core of any such plan. Regarding this point, General Cadorna (Cadorna 1921, I, pp. 23–24) stated:

> The operation plan must be developed before coming to war and is determined by: one's own available forces, those of one's enemy, and one's strategic goals. However, there is an element of paramount importance in determining the operation plan that is constant in any case, that is the terrain, which determines the extent of the obstacle in the case of an offensive war, and the viability that establishes the logistical possibility to overcome this obstacle with a given force.

General Porro (1854–1939), vice-president of the Italian Geographical Society and second in command in the Royal Italian Army during part of WWI, is considered one of the greatest scholars of military geography in Italy. He defines military geography assessment as the determination of the significance and influence of geography on the application of military power (Porro 1898). According to General Porro, military geography assessment can be done theoretically within certain limits. Such an approach concerns those "...few principles regarding the generic functions of different geographic elements in war operations" (Porro 1898, p. 30). He includes the following characteristics that should be considered in theoretical analysis of geographic elements: intrinsic conditions, position in the battlefield, relations with other geographical elements, and orientation with respect to military action. In addition to theoretical analysis, the field of applied military geography provides further practical specifications for these functions. This applied analysis, as a result of evaluating different hypotheses of war, introduces the appreciation of positive data, such as the intensity of the forces in action (see the "eclectic approach" in Porro (1898, p. 31).

Chapter 3 ("The Operational Design") in Cadorna (1921), *The War at the Italian Front*, describes the concepts that were the foundation for Italian operational planning. In the first part of this chapter (pp. 85–87), General Cadorna briefly mentioned the defensive and offensive significance that he discerned for each of the three zones of operation he defined for the Austro-Hungarian theater of war. We consider this part of General Cadorna's text as a synthesis of his theoretical military geography assessment for this theater of war. Later on in this chapter, an evaluation of different hypothetical situations of war is also included. Those regarding the Cadore are on p. 90. On p. 91, one can find a summary of the operational design he finally adopted, which could be considered as the result of the application of his strategic concepts and his appreciation of the forces' intensity to the geography's military significances derived from the theoretical military geography assessment.

General Cadorna divided the theater of war in three zones of operation, the Central Zone of operation included the territory relevant to the Cadore offensive (Fig. 6.1). In his theoretical military geography assessment, describing the Central Zone of operation, General Cadorna highlighted that it formed a salient into enemy territory, whose apex was close to an important Austro-Hungarian interior-lateral line, the Puster Valley. "Interior-lateral line" ("linea di arroccamento") refers to a line of communication with the peculiar feature that it develops more or less parallel to a strategic front. Thus, it allows big, fast, and secure deployment of forces from one line of operation to another or from one point of the strategic front to another. Such lines of communication are of great value in areas where there is a lack of communication from one line of operation to another, such as in mountainous theaters of war. These particular features led General Cadorna to infer that the Central Zone's main offensive significance was that, from the apex of the salient in Cadore region, the Royal Italian Army could cut off Austro-Hungarian communications in the Puster Valley. General Cadorna also noted another important value of this zone for the Italian offensive: once the Puster Valley is reached, it would be possible for the 4th Army to contribute to offensives against the Trentino to the west and the Julian Front to the east (Fig. 6.1).

Historical-Geographical Framework of the Area Relevant to the Cadore Offensive

A description of the area relevant to the planned Italian attack on the Puster Valley must start with the valley itself, because it dominates the physiognomy of the Eastern Alps where it lies. The Puster Valley is approximately 90 km long and runs roughly west to east, parallel to the Alpine mountain range. The main towns in the valley are Brunico, Dobbiaco, San Candido, Heinfels, Mittewald, and Lienz (Fig. 6.2).

The Dobbiaco Saddle (Fig. 6.2), topographically almost imperceptible, divides the valley in two parts, separating the headwaters of the Rienza River, which runs west into the Adige River, from the Drau River, which runs east into the Danube River. The current border between Italy and Austria, which crosses the valley transversally, is very near to and east of the saddle. The western Puster Valley is currently in Italy, whereas the eastern part is in Austria, but until the end of WWI, the entire Puster Valley was part of the Austro-Hungarian Empire. The border which existed before WWI was established in 1866, after the 3rd War of Italian Independence. This border bulged into Italian territory in its western part: the bulge was called the "Trentino salient" (Fig. 6.1). On the eastern base of this salient, the border passed through the northern part of the Dolomites, at one point only 13 km in a straight line from the Puster Valley. The many peaks and deep valleys of the Dolomites determined abrupt changes in elevation along this part of the border. Eastward, the border reached the Carnic Ridge (or the Carnic Alps; Fig. 6.1) where it ran for about 100 km with only minor changes in elevation (Fig. 6.2).

Geology and Tectonics

The Alps are traditionally comprised of the Helvetic, Penninic, Austroalpine, and South Alpine superunits. The physical origin of the Puster Valley can be attributed mainly to the presence of a major tectonic lineament, the Periadriatic Line (Pustertal and Gailtal Lines); this lineament continues into the Gailtal Valley east of the current border between Italy and Austria (Fig. 6.3; Janoschek and Matura 1980). The Periadriatic Line plays a fundamental role in the structural framework of the Alps, separating the Alpine orogeny into Austroalpine nappes and Southern Alps. The Periadriatic Line (along with the related lineaments) is related to the post-collisional deformations of the Alpine chain, which generated a dextral transpression between the Adriatic subplate and the European foreland (Schmid et al. 1989).

The Southern Alps belt is characterized by intense back-thrusting, forming a south-verging orogenic structure (Africa-verging belt) and facing the tectonic polar-

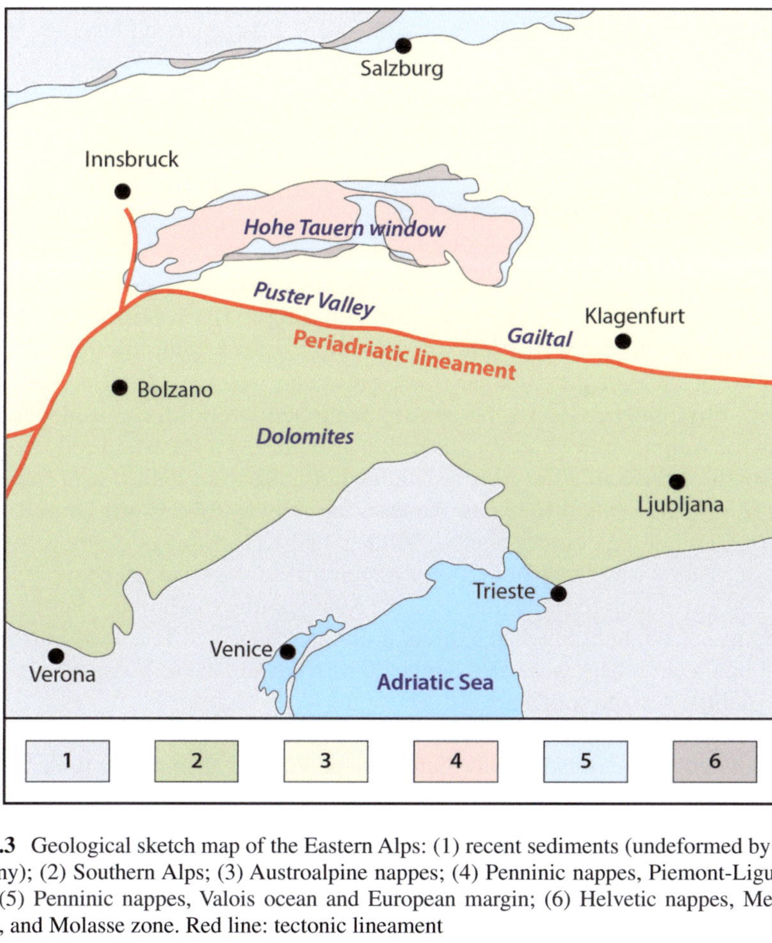

Fig. 6.3 Geological sketch map of the Eastern Alps: (1) recent sediments (undeformed by Alpine orogeny); (2) Southern Alps; (3) Austroalpine nappes; (4) Penninic nappes, Piemont-Liguria terrane; (5) Penninic nappes, Valois ocean and European margin; (6) Helvetic nappes, Mesozoic, flysch, and Molasse zone. Red line: tectonic lineament

ity of the north-verging Northern Alpine chain (Europe-verging orogenic chain), located north of the Periadriatic Line (Castellarin et al. 2003), where Helvetic, Penninic, and Austroalpine nappes lie.

The Austroalpine domain comes in contact with the Southern Alps along the western part of the Puster Valley and is present on both sides of the valley in the east (Dal Piaz et al. 2003). The Penninic tectonic domain, in which the Tauern Alps are located, lies about 30 kilometers north of the Puster Valley (Fig. 6.3).

The Southern Alps run along the southern side of the Gailtal Valley and south of the western Puster Valley (Fig. 6.3). On the southern side of the Gailtal Valley, a long ridge called the Carnic Crest (or the Carnic Alps) runs (Figs. 6.2 and 6.3), where the border between Italy and Austria has settled since 1866.

The succession in the Southern Alps mainly includes late Permian volcanics and siliciclastics and Mesozoic carbonate rocks, sitting on a Ercinian metamorphic basement. In the southwest the Southern Alps disappear below recent sediments of the Po basin that are lying discordant on top of them. The Dolomites, formed of the late Carboniferous to Tertiary succession, extends to the south. These mountains are characterized by groups of high peaks with steep cliffs separated by deep, narrow valleys joined by saddles. Following those valleys southward eventually leads, by a labyrinthine path, to the Veneto Plain.

Transportation Infrastructure

At the time of WWI, the Puster Valley had a railway line that was part of the extended Austro-Hungarian railway system (Figs. 6.1 and 6.2). The western station in the valley was Fortezza, from which it was possible to travel south toward Bolzano or to turn north through the Brenner Pass, eventually reaching the Inn Valley and Germany. From Bolzano, it was possible to continue southward, connecting to Trent through the Adige Valley (Fig. 6.1). Bolzano also offered a westward railway route, going south of the Sarentine Alps to Malles in the Venosta Valley, near the Swiss border (Fig. 6.1). Beyond Lienz, to the east, the railway follows the Drau River to Marburg, through the cities of Spittal, Villach, and Klagenfurt in southern Austria. The Puster Valley was thus part of a very important east-west line of communication nearly 500 km in length, extending from the Swiss border to Marburg. Furthermore, from Spittal, the Tauern line had a direct train that connected Trieste to Salzburg in just 9.5 hours, and there were also connections from Villach to Vienna in the north and to Ljubljana in the southeast.

The density of the railway network in the Italian-controlled areas of this region was much less. The closest railway station to the Puster Valley on the Italian side of the border was the Pieve di Cadore terminal station (Fig. 6.2). From Pieve di Cadore, the Italian line ran for more than 100 km, mainly along the Piave Valley, until reaching its only connection, with the Veneto Plain railway system at its far southern end. This rail line had a low capacity compared to the Austro-Hungarian rail line of the Puster Valley.

Table 6.1 Roadway itineraries from Pieve di Cadore to the Puster Valley finished in the Sesto Valley, Landro Valley, or Badia Valley. Each of these lines of penetration was controlled by an Austro-Hungarian "sbarramenti." Elevations are shown in parentheses

Roadway itineraries from Pieve di Cadore (878 m.a.s.l.) to the Puster Valley	Distance to the Austro-Hungarian "sbarramenti"	Distance to the Puster Valley
Santo Stefano di Cadore (908 m) in the *Piave Valley*–Montecroce di Comelico Pass (1636 m) in the top of the *Padola Valley*–*Sesto Valley*–San Candido (1175 m) in the *Puster Valley*	49 km	65 km
Auronzo di Cadore (866 m) and Misurina (1752 m) in the *Ansiei Valley*–Carbonin (1440 m)–Lago di Landro (1406 m) in the *Landro Valley*–Dobbiaco (1256 m) in the *Puster Valley*	49 km	64 km
Cortina d'Ampezzo (1211 m) in the *Boite Valley*–Tre Croci Pass (1805 m)–Misurina (1752 m) in the *Ansiei Valley*–Carbonin (1440 m)–Lago di Landro (1406 m) in the *Landro Valley*–Dobbiaco (1256 m) in the *Puster Valley*	52 km	66 km
Cortina d'Ampezzo (1211 m) and Fiames (1293 m) in the *Boite Valley*–Cimabanche Pass (1529 m)–Lago di Landro (1406 m) in the *Landro Valley*–Dobbiaco (1256 m) in the *Puster Valley*	50 km	63 km
Cortina d'Ampezzo (1211 m) in the *Boite Valley*–Falzarego Pass (2117 m) in the Costeana Valley–Val Parola Pass (2192 m)–Badia (1330 m) in the Badia Valley–Brunico (838 m) in the *Puster Valley*	50 km	94 km

Roadway infrastructure in Cadore was not better for the Italians than the railway infrastructure. The journeys from Pieve di Cadore to the Puster Valley required using roads that often took hairpin turns, included major fluctuations in elevation, and that ran through tortuous valleys that also involved going over mountain passes. Even though the linear distance between the two places is around 30 km, road journeys were long and demanding and covered at least twice that distance. Itineraries from Pieve di Cadore finished on one of either the Sesto Valley, Landro Valley, or Badia Valley, each of which led into the Puster Valley (Table 6.1).

Theoretical Military Geography Considerations Relevant to the Cadore Offensive

Studying the area of the Cadore offensive from a military geographical point of view requires an approach considering military geography theory developed prior to WWI. The main reference for our analysis is the "Study Guide for Military Geography" (Porro 1898). This work is a summary of the lessons in military geography that General Porro taught at the Italian Army School of War. It is still the foundation for the study of military geography in Italy today.

General Porro's first consideration that is important to highlight regards the mountainous terrain. This kind of terrain, considered in a general sense, is the least suitable for carrying out any quick and decisive operations using large numbers of men. This is a fundamental principle of conventional warfare. The reasons for this unsuitability are the harshness of the terrain, the complex climatic conditions, and the lack of communication facilities, built-up areas, and resources in such areas (Porro 1898).

In most cases of war in mountainous terrain, valleys represent the geographic elements with the greatest strategic, logistical, and tactical importance (Porro 1898). For General Porro, the theoretical study of the military functions of mountain valleys must focus on their shape and size and on the mountain range in which they are located, for example, the orientation of the valley within the mountain range and the spatial relations between the valley and other geographical features in the mountain range (Porro 1898). Such theoretical study must be considered with respect to scale to determine their strategic, operational, and/or tactical significance.

Small Scale

Professor D.W. Johnson, who had long been interested in the military aspects of geography during and after WWI, provided a clear overview of the major geographical features of the Eastern Alps and their topologic relations from strategic and logistical perspectives (Johnson 1917 and 1921). Through this maze of rugged mountains, Johnson revealed that the Puster Valley is part of a series of parallel glacial-tectonic trenches of the greatest strategic importance in warfare. These trenches are, from north to south, the Inn corridor, the Venosta Valley-Puster Valley corridor, the Noce corridor, and the Valsugana corridor (Fig. 6.1).

These corridors are both strategically and logistically important; extending in an east-west direction, they control all longitudinal movements inside the Alpine range. Their importance is also due to the extremely high mountain crests which separate the corridors among themselves as well as separating them from the plains to the south. These valleys thus each provide secure lines of communications.

The tenuous connections between the corridors and between the corridors and the plains are comprised of deep, narrow, north-south valleys and, in some cases, just barely accessible mountain passes. The most noteworthy of these north-south connections is a continuous corridor formed by the valleys of the Sill River, which flows north, and the Eisack River, which flows south from the Brenner Pass. This main north-south corridor (Fig. 6.1) was not only morphologically important for military activity but was also already densely packed with transportation infrastructure: it was a crucial communication path because it was the rail and road connection between Italy and Germany. It also crosses and connects all of the east-west lateral corridors that Johnson identified (Johnson 1921).

Another factor that increases the Puster Valley corridor's value as a line of communication is that it runs nearly parallel to the border between Italy and Austria-Hungary. Thus, it runs perpendicular to the direction of the two countries' military advances toward each other. For the Hapsburgs, the Puster Valley constituted an interior-lateral line through which troops and supplies would be rapidly and safely shifted from one point to another along the battle front once fighting began.

Medium Scale

At a regional scale, however, one can perceive the vulnerability of the Puster Valley. The wide and high western Tauern Mountains completely cut the Puster Valley off from communication with the Inn corridor to the north and thus preclude any tactical support for the Austro-Hungarian defense of the valley from that direction. Furthermore, if the Royal Italian Army had reached the valley, they would have split the Austro-Hungarian forces, enlarged the defensive front, and, most importantly, considerably lengthened the Austro-Hungarian lines of communication between the Trentino and the Carnic fronts of war. One can easily conclude that this would strongly impact not only the Italian front but the entire Austro-Hungarian theater of war, as a successful Italian assault on the Puster Valley would have overloaded the rest of the Austro-Hungarian railway system.

South, from where the Italian attack had to take place, the Puster Valley presented two different tactical situations. East of the Dobbiaco Saddle, the Puster Valley could be considered naturally well defended: mountain masses sit between the Puster Valley and the smaller but parallel Gailtal Valley. Another chain of mountains south of the Gailtal Valley, the Carnic Ridge, constitutes a further topographical barrier (Fig. 6.2). West of the Dobbiaco Saddle, the Puster Valley joins the three small, narrow, north-south valleys, the Sesto, Landro, and Badia Valleys (Fig. 6.2), which lead into roads that reach the most upstream area of the Piave Valley. Despite the fact that this upstream region of the Piave Valley was the most important Italian interior-lateral line inside the Central Zone of operations, its strategic and logistical importance was rather limited compared to that of the Puster Valley. The Piave Valley was narrow, extends from southwest to northeast, and contained very little infrastructure. These problems detracted from its potential as a communication corridor and as a staging area for soldiers.

Large Scale

The itineraries which, beginning from Pieve di Cadore, led to the Austro-Hungarian "sbarramenti" and then to the Puster Valley (Table 6.1) represented a major logistical challenge for the Royal Italian Army. Furthermore, from the tactical point of view, they were very unfavorable for any Italian offensive. First of all, the morphology of these valleys, which in some parts were incredibly narrow, made any numerical superiority of an attacking army utterly worthless (Cadorna 1950). Second, the presence of the Austro-Hungarian fortifications, the "sbarramenti," would make any advance through the valleys very difficult. Last but certainly not least, it can be assumed that the planimetric configuration of these lines of penetration, separated by mountain masses, would prevent any significant tactical cooperation between the columns advancing along these. For instance, a column operating against the "sbarramenti" in the Sesto Valley would have been completely separated from those attacking the "sbarramenti" in the Landro and Badia Valleys (Fig. 6.2). Moreover, those attacking the "sbarramenti" in the Landro and Badia Valleys would have operated with only slim possibility of communication between them.

As for the Austro-Hungarian defenders, the maximum distance along the Puster Valley between the western entrance, Sesto Valley, and the eastern entrance, Badia Valley, is less than 40 km. Combined with the Puster Valley's solid transport infrastructure (see above), Austro-Hungarian troops would have been able to travel quickly, providing a flexible defense at all the various entrances to the Puster Valley.

However, the Royal Italian Army would have some advantages if it could have reached the Puster Valley. Even though long valleys offered to protract defense at different points (Porro 1898), the Puster Valley's general wide cross-section and low-sloping sides (Tolomei 1910) would have afforded the Italian troops an offensive advantage in terms of deployment from a tactical point of view.

Logistical Considerations of the Cadore Offensive

Logistic infrastructure, as the terrain, is an element of paramount importance in determining operation plans, since no matter the plan, it remains the same (Cadorna 1921; see section "Cadorna's Theoretical Military Geography Assessment of the Central Zone of Operations (Cadore)"). Because logistics relies on concrete and verifiable foundations, logistics history allows for objective judgments on the feasibility of hypothetical or actual operation plans (Botti 1991). From this point of view, logistical judgments about the feasibility of the Cadore offensive constitute a valuable complement to our theoretical military geographic considerations, as they can help us to clarify objectively the extent of the geographical obstacles and the difficulties related to overcoming them.

Senior Italian officers during WWI offered their own judgments about the logistical feasibility of the Cadore offensive, both during and after the war. General A. Gatti argued (Gatti 1929) that the Austro-Hungarians had far more logistical capabilities for defending Cadore than the Italians had for their attack, implying that General Cadorna's operational plans overestimated the logistical capacities of the Royal Italian Army.

Gen. Gatti's opinion was based on the fact that the transport capacity of the Italian railway line was such that it would have taken an entire month to gather the necessary troops for an army corps at the Pieve di Cadore terminal station. This problem was compounded by the length and difficulty of the roads north from Pieve di Cadore to Austro-Hungarian territory (Gatti 1929).

General G. Liuzzi, the logistical staff officer of the 4th Army, gave a particularly clear assessment of the Italian transportation problem, writing in June 1915 that:

> If the 4th Army reached Innichen (San Candido) in September, should it operate towards the east or west while continuing to rely upon the railroad of the Piave Valley, with its terminal station in Calalzo (Pieve di Cadore Station), and on the ordinary streets of M. Croce Comelico and Schulderbach for its supplies and evacuations, by late autumn, and perhaps before, it will be inevitably separated from its base and from its Logistical Command that, innocent and pure, will have to declare collapse. (Liuzzi 1922, p. 51)

Results

The offensive military significance General Cadorna assigned to the Central Zone in his theoretical military geography assessment (section "Cadorna's Theoretical Military Geography Assessment of the Central Zone of Operations (Cadore)") can be seen from the perspective of the theoretical military geography considerations (section "Theoretical Military Geography Considerations Relevant to the Cadore Offensive") and logistical considerations (section "Logistical Considerations of the Cadore Offensive") provided in this paper:

- The Puster Valley had the characteristics to become an important Austro-Hungarian interior-lateral line because of its length along and proximity to the border between Austria-Hungary and Italy and its significant towns and infrastructure but, most of all, because of its importance for the Austro-Hungarian communications system. From a local perspective, the Puster Valley guaranteed the flexibility of troop movements along the border. At the level of the theater of war, it linked Trentino front, which was connected with Innsbruck in Austria and with Germany, with the Julian and Carso fronts which were connected with the Serbian theater of war, and with the heart of the Austrian monarchy.
- The offensive significance that General Cadorna ascribed to the Central Zone for conducting an attack toward the Puster Valley is considerably diminished when compared with our theoretical military geography considerations and with the collected judgments on logistics. Despite the proximity to the Puster Valley, which General Cadorna highlighted, this offensive significance decreased because there was no reliable interior-lateral line inside Italian territory, given the low level of communication and infrastructure development of the inner Piave Valley. The morphological and planimetric characteristics along the Italian lines of penetration from the Piave Valley to the Puster Valley offered further tactical difficulties.
- The results of our theoretical considerations accord, in part, with those of General Cadorna with regard to the offensive significance of the Puster Valley as a potential line of penetration into Austro-Hungarian territory. In particular, the Tauern Mountains to the north would have covered the flank of the advancing Italian

column, and the width of the valley would have provided possibilities for force deployment and maneuverability simply not possible in other valleys in the Eastern Alps. However, the considerations regarding logistics that we have examined (section "Logistical Considerations of the Cadore Offensive") indicate such movement, both westward and eastward along the Puster Valley, would have been inadequate and thus probably doomed to failure.

Discussion

Although most authors who have written about Italian military operations during WWI noted the unfavorable geography of the Cadore, where the first Italian offensive occurred, no one has yet explored this topic in depth via a study of military geography. Furthermore, many authors analyzed and criticized General Cadorna's planning (e.g., his operation plan and operational design) without considering his assessment of the different elements that were the basis for his recommendations, e.g., terrain and logistical infrastructure. From this perspective, we believe this study concerned with military geography represents an original approach and a solid point of departure from which to study operational planning, which in turn could lead to a better understanding of historic warfare events in a given territory.

However, we have to note a difficulty in carrying out the methodology proposed in our study case: General Cadorna (1921) only devoted a "brief mention" to the defensive and offensive significance in the text that we identified as his theoretical military geography assessment. Indeed, although General Cadorna very clearly stated the primary military significance he attributed to the various zones of operation, there is no extended presentation of his theoretical military geography assessment. Such a presentation would have allowed for a more detailed study and further conclusions. Additional considerations of military geography from General Cadorna could be gathered from the operation plan and the assessments he made for different hypotheses of war, including numerous considerations on tactics and logistics. However, as previously mentioned, this paper did not consider references beyond those identified as the theoretical assessment.

Regarding the results of this paper (section "Results"), we suggest two hypotheses that could explain the differences in terms of the military significance of the Cadore offensive's geography that we identified between General Cadorna's assessment (Section "Cadorna's Theoretical Military Geography Assessment of the Central Zone of Operations (Cadore)") and the considerations that we provided in this paper (sections "Theoretical Military Geography Considerations Relevant to the Cadore Offensive" and "Logistical Considerations of the Cadore Offensive"). First, the military significance that we identified in General Cadorna's chapter "The Operational Design" as being theoretical could have been influenced by considerations unrelated to theoretical military geography. For example, it could have been affected by strategic issues related to different hypotheses of war, which General

Cadorna developed in the second part of the chapter (e.g., the impossibility of conducting an offensive on the Julian and Carso fronts at the beginning of the war).

The second hypothesis is that General Cadorna may have underestimated the tactical and logistical problems that an offensive from Cadore would have to face. This is suggested by the lack of any reference to tactical or logistical issues in this part of General Cadorna's text, where the main military significance of the zones of operation was identified. In this regard, however, it is worth noting that, as stated above, what we identified as General Cadorna's theoretical military geography assessment refers to just a "brief mention." Elsewhere, General Cadorna (1921) presented a theoretical assessment of the general tactical conditions along the entire Italian front, and, in other parts of the same book, he also assessed the logistical and tactical impediments of a hypothetical mass attack on the Trentino salient and the entrenched camp of Trent.

Our last hypothesis is supported in the literature, with respect to the logistical difficulties, by General F. Botti, who stated that logistics is "…perhaps the weakest and less realistic side of General Cadorna's operational design, hitherto strangely ignored by historical critique…" (Botti 1991, p. 680). Isnenghi and Rochat (2008) considered that General Cadorna did not fully take into account the geographic obstacles. Regarding the Cadore offensive, they stated that the fact that General Cadorna planned this offensive showed that he had not yet realized how difficult it was to conduct an offensive, especially in mountainous terrain (Isnenghi and Rochat 2008).

Conclusions

The military geographic approach we presented has allowed us to focus on analyzing physical geographical features, which are often overlooked in military-historiographical analysis in favor of other aspects, such as geopolitical, organizational, economic, or social issues. In particular, the approach proposed in this paper enabled us to analyze General Cadorna's theoretical assessment of the Cadore's offensive geography, which was the basis of his operation plan. The results of the analysis can be summarized as follows:

1. Our theoretical military geography considerations accord with General Cadorna's belief in the importance of the Puster Valley as an Austro-Hungarian interior-lateral line.
2. The potential of the Central Zone for conducting an offensive aimed at reaching the Puster Valley and of this valley as a line of penetration into Austro-Hungarian territory, which General Cadorna identified, is significantly diminished when compared with our theoretical military geography considerations and with the collected judgments on logistics.

These results accord with the idea proposed by other authors (Botti 1991; Isnenghi and Rochat 2008) that General Cadorna's evaluation of the difficulties imposed by the Cadore offensive's geography was not entirely accurate. Future studies in this regard should examine the military geography assessments of other military figures and military scholars. In addition, the use of geographical information system (GIS) has been demonstrated to be a valuable tool for the study of the relationship between physical geography and military history (e.g., Guth 2011; Bondesan et al. 2013, Rua et al. 2013; Vergara et al. 2017). In this sense, the introduction of quantitative data might contribute to a clearer understanding of the influence and significance of geography in the failure of the Cadore offensive.

Acknowledgments This work stems from Mauricio Nicolas Vergara's PhD dissertation, which was supervised by Professor Aldino Bondesan, and received funding from the Foundation for University and High Culture in the Province of Belluno and the MIUR ex60% funds (Professor Bondesan). The authors want to acknowledge in particular "reviewer 2" who suggested many improvements, as well as Margarita Petit and Scribendi (editor EM825) for the language revision.

Bibliography

Berti, A. (1982). *Guerra in Ampezzo e Cadore (1915–1917)*. Milano: Arcana.

Bondesan, A., Furlani, S., Vergara, M. N., et al. (2013). Geomorphology of the El Alamein Battlefield (Southern Front, Egypt). *Journal of Maps, 9*, 532–541.

Botti, F. (1991). *La logistica dell'esercito italiano (1831–1981)*. Roma: Ufficio Storico Stato Maggiore dell'Esercito.

Cadorna, L. (1921). *La guerra alla fronte italiana: fino all'arresto sulla linea della Piave e del Grappa (24 Maggio 1915 – 9 Novembre 1917)*. Milano: Treves.

Cadorna, L. (1950). *Pagine polemiche*. Milano: Garzanti.

Castellarin, A., Cantelli, L., Bertelli, L., Borrini, D., Fantoni, R., Sella, M., Nicolich, R., Mazzotti, A., Selli, L., Bernabini, M., Lammerer, B., Lueschen, E., Gebrande, H., Millhan, K., Grassl, H., Neubauer, F., Oncken, O., & Stiller, M. (2003). The TRANSALP seismic profile and the CROP 1A sub-project. *Mem. Descr. Carta Geol. d'It., 62*, 107–126.

Dal Piaz, G. V., Bistacchi, A., & Massironi, M. (2003). Geological outline of the Alps. *Episodes, 26*(3), 175.

Di Martino, B., & Cappellano, F. (2007). *La Grande Guerra sul fronte dolomitico*. Novale: Rossato.

Gatti, A. (1929). *Uomini e folle di guerra*. Milano: Mondadori.

Guth, P. L. (2011). Military applied geomorphological mapping: Normandy case study. *Developments in Earth Surface Processes, 15*, 577–588.

Isnenghi, M., & Rochat, G. (2008). *La Grande guerra (1914–1918)*. Bologna: Il Mulino.

Janoschek, W. R., & Matura, A. (1980). *Outline of the geology of Austria*. Wien: Verlag der Geologischen Bundesanstalt.

Johnson, D. W. (1917). *Topography and strategy in war*. New York: Henry Holt.

Johnson, D. W. (1921). *Battlefields of the world war*. New York: Oxford University Press.

Liuzzi, G. (1922). *Ricordi e pensieri di un ex-intendente d'armata*. Roma: Stabilimento Poligrafico per l'Amministrazione della Guerra.

Nava, L. (1922). *Operazioni militari della 4° Armata nei primi quattro mesi della campagna di guerra 1915*. Cherasco: Raselli.

Pieri, P. (1998). *La prima guerra mondiale (1914–1918)*. Udine: Gaspari.

Porro, C. (1898). *Guida allo studio della geografia militare*. Torino: UTET.

Romolotti, G. (1978). *Il 24 maggio 1915*. Milano: Domus.

Rua, H., Gonçalves, A. B., & Figueiredo, R. (2013). Assessment of the Lines of Torres Vedras defensive system with visibility analysis. *Journal of Archaeological Science, 40*, 2113–2123. https://doi.org/10.1016/j.jas.2012.12.012.

Schmid, S. M., Aebli, H. R., Heller, F., & Zingg, A. (1989). The role of the Periadriatic line in the tectonic evolution of the Alps. In M. P. Coward, D. Dietrich, & R. G. Park (Eds.), *Alpine tectonics* (Special Publications) (Vol. 45, pp. 153–171). London: Geological Society.

Tolomei, E. (1910). Particolare della linea Spartiacque sul Passo di Toblaco. *Archivio per l'Alto Adige, 5*, 255.

Vergara, M. N., Bondesan, A., & Ferrarese, F. (2017). GIS analysis of the trafficability determined by slope in the eastern Tyrol front (WWI, Eastern Alps): A military history reading. *Cartography and Geographic Information Science.* https://doi.org/10.1080/15230406.2017.1399828.

Chapter 7
"Trenches in the Ocean": Geography and the Russian Defense of the Baltic Sea, 1912–1917

Michael B. Barrett

With war on the horizon since the Bosnian Crisis of 1908, treaty obligations made it probable that Russia faced two enemies to her west: Germany and Austria-Hungary. The former was the more dangerous militarily, for her army and industrial base were second to none, while the Habsburg forces enjoyed mediocre leadership, had logistical difficulties, and were riven by national animosities. Russia's real arch-enemy was Austria-Hungary. The Danube Empire's interests in the Balkans could not tolerate any further Russian incursions, and Pan-Slavism dictated Russian protection of faith and culture in that stormy area. Ironically, Russia did not really have a grievance with Germany. What propelled the two toward war was Russia's treaty commitment to France to attack Germany. Russia's political interests really dictated holding against the Germans and attacking the Austrians.

Unfortunately for Russia, geography worked against her. Geographic or natural barriers to a German advance simply did not exist. The only serious impediment between Berlin and St. Petersburg was distance. To the south, the Austrians held the high ground. From the juncture of the German, Austrian, and Russian borders in Upper Silesia, a chain of mountains (the Tatra, Beskiden, and Carpathians) formed an impenetrable wall around the Hungarian heartland.

From the sea, matters were worse. The Baltic and Black Seas were bodies of water whose outlets were controlled by neutrals or potential foes. Both seas extended deep into Russia and could not be ignored as potential avenues of approach. In the Black Sea, the Russians faced a threat from Turkey, but here they had a reasonable chance to come out ahead given Turkey's relative naval weakness (Halpern 1994). In the north, most of Russia's foreign trade ran through the Baltic. She had to defend a very long coastline on the Baltic, and her capital sat at its eastern end, vulnerable to naval assault. The Baltic constituted Russia's most important naval theater.

M. B. Barrett (✉)
Department of History, The Citadel, Charleston, SC, USA
e-mail: barrettm@citadel.edu

P. L. Guth (ed.), *Military Geoscience*, Advances in Military Geosciences,
https://doi.org/10.1007/978-3-030-32173-4_7

The three neutral nations located at the western entrance to the Baltic, Denmark, Norway, and Sweden controlled entry and exit from this huge inland sea. In addition, the channels from the North to the Baltic Sea were so constricted and lay so close to Germany that the Germans could interdict or even close them any time they desired, which they in fact did when the war broke out. The Baltic Commander in Chief, Prince Henry of Prussia, had both Danish belts mined, although the Danes later added plenty of their own mines (Halpern 1994) (Fig. 7.1).

The narrow entry and exit to the Baltic and its proximity to Germany effectively cut off Russia from her most important allies, France and England. The Germans cleverly sidestepped the issue of neutral nations controlling the entrance to the Baltic Sea by constructing a canal, the Kaiser Wilhelm or Kiel Canal, across their section of Jutland. Improvements to the canal completed in the summer of 1914 allowed them to move even their largest capital ships from the Baltic to the North Sea or vice versa in complete safety within a matter of hours. Finally, even the formidable climate of

Fig. 7.1 Baltic approaches

the Baltic hampered the Russians more than her neighboring rivals. Significant sections of the northern and northeastern Baltic froze each winter, primarily the Gulfs of Bothnia, Finland, and Riga, all located in Russian waters (Fig. 7.2).

For all intents, the winter freeze shut down the Russian Navy from January to April. Only Libau, then at the farthest extent of the Russian western frontier, largely remained ice-free all winter. Alas, it was so remote and exposed the Russians recognized they could not hold it and planned to abandon it in the event of hostilities (Pavlovich 1979).

The poor hand dealt by geography to the Russians in the Baltic was immeasurably worsened when Russia lost her Baltic Fleet to the Japanese in 1905 at the Tsushima Straits between Korea and Japan. While that decision settled the issue of dominance in Russia's third naval theater in favor of Japan and relieved Russia of having to maintain a Pacific Fleet, the catastrophe left both her Baltic coastline and capital sitting at its eastern end undefended. In 1906 the Baltic Fleet consisted of only three battleships, one obsolete and two about to be, along with a few modern cruisers. It could not expect to take on the German High Sea Fleet, which possessed alone ten battleships with four of the new dreadnought-type under construction (Pavlovich 1979).

Russia had the space to buy the time to try to thwart any advance by land from East Prussia. Any German army attempting to take St. Petersburg had to slog its way north through nearly 800 kilometers of enemy territory. On the other hand, without any realistic naval defenses in the Baltic, a German fleet could land troops with impunity just kilometers from St. Petersburg. The Russian general staff estimated it would take 2 weeks to mobilize and assemble sufficient forces near St. Petersburg capable of defending it. With German ports within 30 hours of sailing time of St. Petersburg and nothing in the way to stop them, the Russian army could not protect the capital. Only a navy could, and for all intents and purposes, Russia did not have one.

In the aftermath of Tsushima, major reforms should have begun, but political considerations, finances, and the tsar's reluctance to permit necessary administrative reforms made sure nothing happened with alacrity. New ship construction necessarily constituted a long-range undertaking, but the capital's defenses could not wait. In April 1906, Navy Minister Admiral Alexei Alexseyevich Birilev (1844–1915) created a navy general staff to solve strategic problems, plan for the rebuilding of the fleet, and manage mobilization. Given the preponderance of young and reform-minded officers on this new staff, its recommendations were hardly surprising. The staff's first significant study called for creating a new Baltic Fleet capable of defending the Gulf of Finland and taking the fight to the Germans. While such views found resonance in the Foreign Ministry and numerous interest groups in an excited public, resistance formed in the army general staff as well as from a few practical-minded officers in the Admiralty (Podsoblyaev 2002).

Senior army officers regarded securing the coastline as the army's mission that could be accomplished by dispatching troops to threatened areas. They had no truck with an Alfred Thayer Mahan-like navy taking command of the seas, and, above all, draining money from the land service's own rebuilding program. This did not sit

Fig. 7.2 Baltic theater

well with the new Navy Minister, I.M. Dikov, who got the tsar to agree to a "small program" of construction that would permit the fleet to defend the Gulf of Finland. Despite winning over the tsar, this program fell victim to politics and financial setbacks, taking 4 years just to see its start (RGAVMF, Fond 417, File 4191; Mitchell 1974).

Moreover, not all naval officers saw an immediate construction program as the solution to defending the capital region. Led by the navy's chief designer, Admiral Nikolai N. Beklemishev, a coterie of officers insisted that protecting the capital from an invasion was the navy's primary mission and, given its weakness, along with the time needed to procure ships, using mines and submarines offered great promise in defending against a seaborne invasion (Podsoblyaev 2002).

The use of minefields and submarines as a main line of defense proved anathema to the Young Turks of the navy general staff, but the task of defending the Gulf of Finland was immediate, and the practical-minded men won out. The navy general staff wrote its first plan for defending the Baltic Sea in 1907. The staff assumed the enemy would be Germany and Sweden, both of whom would land troops on the north coast of the Gulf of Finland in order to attack in the direction of St. Petersburg. As Finn historian Pertti Luntinen wrote, "compounding this bleak assessment, the Russians accepted as a dogma of faith that the disloyal Finns longingly awaited such an invasion as the fanfare to Finnish independence. They anticipated that the Finns would assist the invaders and sabotage the Russian efforts at defense, and they planned to introduce martial law immediately upon mobilization (Luntinen 1997)." Neither the landings nor an advance could be prevented or halted in the western portion of the Gulf. Consequently, the chiefs of both the army and navy general staffs agreed that the main task of Russia's armed forces in this area was to defend the Gulf of Finland east of the Gogland Islands. The fleet's mission was to delay for 14 days any enemy advances into the Gulf so ground forces could mobilize and deploy to defense lines on the Kyumen River to the north of St. Petersburg and the river Narva southwest of the capital. To accomplish this mission, the navy general staff came up with a scheme using a combination of minefields and long-range guns (Fig. 7.3).

If mines were set down in sufficient quantity and in a proper pattern, the probability of mines detonating against enemy ships was very high. If the average width or beam of the enemy vessels was 15 meters, mines laid every 45 meters apart have a 33 percent probability of detonating against one of the ships attempting to pass through the line of mines. It would take 22 mines with a separation of 45 meters to cover a kilometer-wide channel. Several rows of mines, offset one behind the other, significantly raise the chances of getting a hit. Four rows of mines with 45 meters of separation, or 88 mines to cover a field a kilometer wide, raise the chances of hitting a passing vessel to 80 percent. Add 2 more rows, another 44 mines, and the probability of getting a hit becomes 90 percent. That probability assumes the ship is attempting to penetrate perpendicular to the rows of mines. If the vessel has to approach at an oblique angle, the chances of hitting a mine become greater (Cowie 1949). The Baltic, and especially the Gulf of Finland, had plenty of shallows and rocky shoals which forced vessel traffic into relatively narrow channels, making the use of mines an attractive strategy.

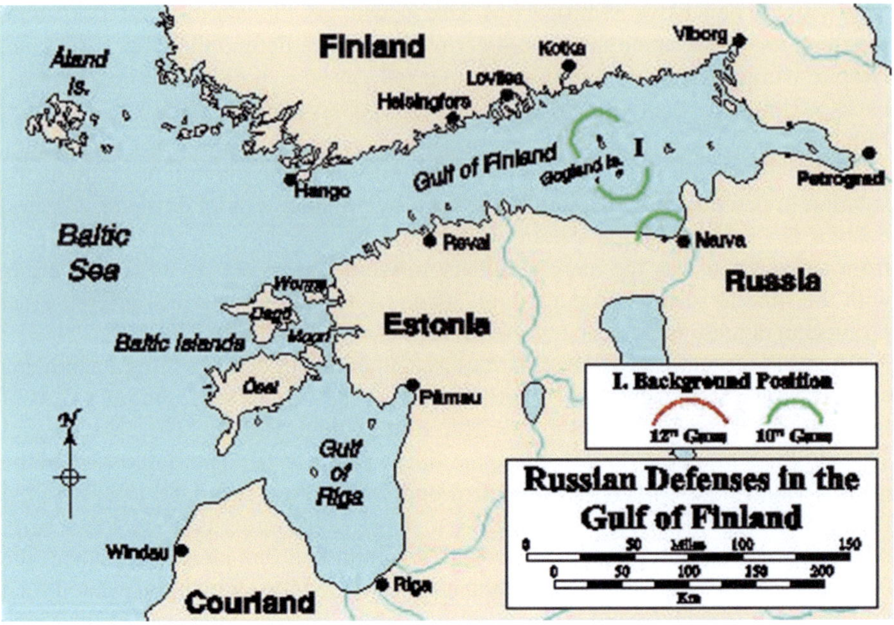

Fig. 7.3 Gogland Islands position

Upon the outbreak of war, the Russian plan called for their fleet to lay mines north and south of Gogland Island. Cruisers would wait at the mouth of the Gulf for the German fleet, and when it arrived, they would retreat to the Gogland Island position covered by their battleship squadron. Destroyers located in the Gulf of Finland between Helsingfors and Kotka (to the east) would lay mines and hinder the enemy flanks, as would submarines. The battleship division would engage the enemy from behind the minefields and hold for as long as possible in conjunction with coastal artillery on Gogland Island (Achkasov et al. 1979; Podsoblyaev 2002).

This plan's weakness was all too obvious: it called for abandoning the two key ports of Reval and Helsingfors which were halfway down the Gulf, and it left the main line of defense too close to the capital. Worse, when tested in 1908, the plan did not work. The Baltic Fleet had too few resources, and poor use was made of them. The bright spot came with the work of the mine division commanded by Rear Admiral Nicholas Ottovich Essen (1860–1915) who sought to enlarge the role of the fleet in the defense of the capital when he became commander of the Baltic Fleet in 1908. Regarded as the ablest Russian naval officer of the World War I and one of the very few naval officers to survive the war with Japan with his reputation intact, Essen believed the Germans would not commit their best ships and the major part of their fleet to operations in the Baltic. He accordingly favored a more aggressive course of embarking on raids to mine German home waters and urged moving the center of gravity for defending the Gulf westward. He further planned to deter Sweden from entering the war by threatening to lay mines outside the Swedish fleet's main base of Karlskrona. If this deterrence failed, his next step was to attack Karlskrona and Stockholm (Achkasov et al. 1979; Podsoblyaev 2002). These plans were too ambitious for both the Baltic Fleet (which had no modern battleships) and the navy general staff which promptly rejected the plan.

Essen did not give up easily. In 1910 he browbeat the navy general staff to add an additional layer or belt of minefields to the defense scheme. He proposed a new line of minefields and artillery much further west, running from Porkkala-Udd in present-day Finland (just west of Helsingfors) to Nargen Island just off the coast of Reval (Podsoblyaev 2002). Essen planned to start his delaying action there, holding out as long as he could before fleeing behind the background defense position centered on the Gogland Islands. In 1912 delivery of new ships strengthened the fleet and allowed Essen to make his stand on the Porkkala-Udd—Reval line of minefields. This new barrier was now designated the main or central defense position, and the fleet was brought forward from Kronstadt to Reval and Helsingfors. The Gulf of Finland was narrower here than on the Gogland Islands meridian, and there were more islands as well scattered about the coastline and the approaches to the harbors, facilitating the placement of shore artillery. Abandoning the Gogland Islands position also meant moving the destroyer force intended to harass the enemy fleet from Kotka on the north shore of the Gulf of Finland east of Helsingfors. Essen moved the destroyer division far to the west behind the line of rocky islets (skerries) running on the north side of the gulf from its mouth at Hango to Helsingfors. He planned to initiate his delaying action at the mouth by deploying his battleships and cruisers, where they would engage the Germans and retreat back behind the central position. Smaller vessels (submarines and destroyers) would dash out from behind the Finnish skerries and harass the Germans. Mine-laying vessels would block the Moon Sound on the eastern side of the Gulf of Riga (see Fig. 7.4) (Halpern 1994; Achkasov et al. 1979; Kozlov 2005). The Russians planned to take advantage of the narrow waters in the western end of the Gulf of Finland which restricted the enemy's maneuverability, forcing him into vast minefields covered by coastal or naval artillery, bringing his advance to a halt. In spite of moving their main line of resistance to the west, the concept of the defense using minefields nonetheless remained unchanged, and it amounted to positional warfare. In fact, the Russians were so conservative that the navy came under the command of the army in time of war. Schedule 19, the Russian war plan, subordinated the Baltic Fleet to the high command (*Stavka*) which placed it under the Sixth Army, the ground force responsible for protecting the capital (Menning 1992) (Figs. 7.4 and 7.5).

On paper, the "Plan of Operations of the Naval Forces of the Baltic Sea in the Event of a European War in 1912" looked good. In reality, it left the initiative with the enemy, and it suffered from material shortages. The Baltic Sea fortresses were in poor shape. The War Department (the Army) had responsibility for them, and they suffered from neglect. The War Department also opposed modernizing them. The controversy naturally delayed the start of construction. Only in 1911 did work begin, and the Admiralty decided to make Reval its main naval base, designating it the "Reval Fortified Zone." In fact, what moved construction along was Essen's successful demand that the fortresses be placed under his command, which occurred in

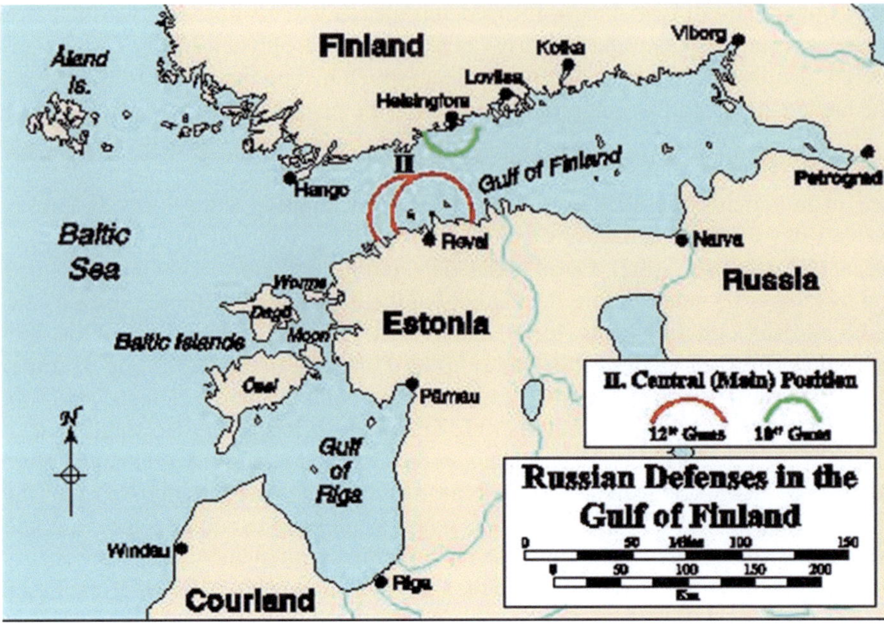

Fig. 7.4 Porkkala-Udd or central defense position

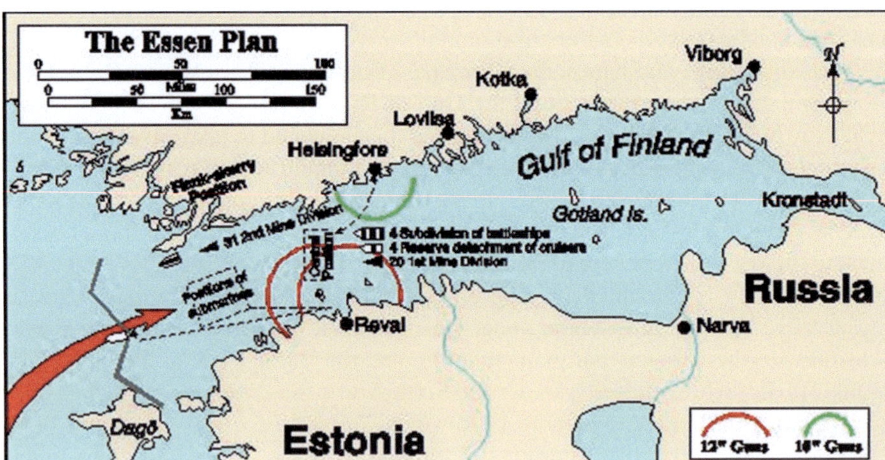

Fig. 7.5 Essen Plan

August 1912. In January 1913, the defense line from the Porkkala-Udd and Reval fortresses became the main defense position of the navy, receiving the name, the "Fortress of Peter the Great."

The 1912 defense plan called for four 14 inch (356 mm) guns in the Surop Islands on the north shore, at Porkkala six guns, and on Makiloto Island, six guns,

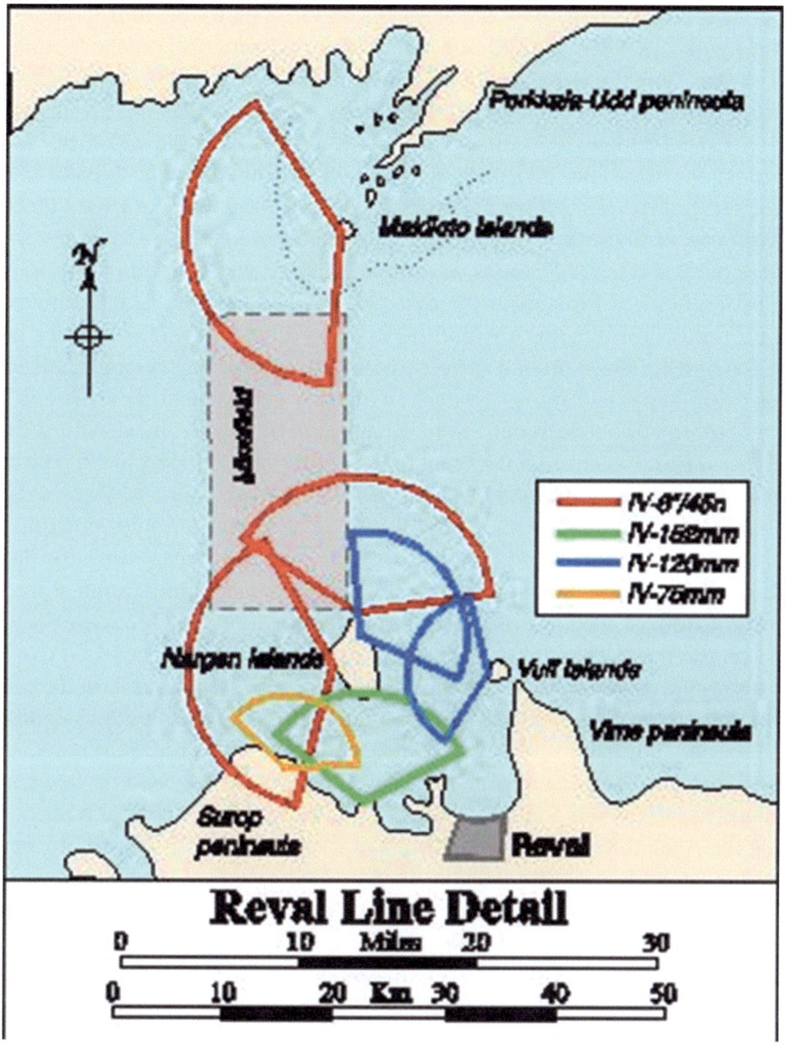

Fig. 7.6 Nargen-Makiloto line

all of 14 inch size. On the south shore of the gulf, Vulf and Nargen Islands guarding the approaches to Reval each received four 12 inch (305 mm) guns to protect the harbor. There were also large howitzers in case the enemy broke through, and some 46 6 inch (152 mm) guns were placed on the various islands to defend passages. Construction on the fortifications and harbors was to finish no later than fall 1917, but delays developed once the war began and the government had to authorize equipment substitutions. At Nargen and Makiloto Islands, the navy authorized three 8 inch (203 mm) batteries of four guns as substitutes for 14 inch and 12 inch guns (Fig. 7.6).

When war broke out in 1914, the Russians had 12 8 inch (203 mm) guns; 4 6 inch (152 mm) guns; 12 7.4 inch (120 mm) guns; and 8 75 mm (2.95 inches) guns in the central defense position. In the Finnish skerries, they had only seven batteries of light artillery (Achkasov et al. 1979). Obviously 12 8 inch guns were a far cry from the planned 16 14 inch guns. Russian industry did not have the capability to make these guns; the Russians had ordered these from England. The blockade of the North Sea meant the guns never arrived. The Russians were also short 12 inch guns, and priority went to the new ships under construction.

Where the Russians did excel was in laying mines, some several thousand at the mouth of the Gulf of Finland in the first few days of the war. The Germans, who kept their focus on England as Admiral Essen predicted, developed a very healthy respect for mines. These proved to be a two-edged sword, however, for current, ice, and corrosion separated mines from their anchor chains, and they drifted where tides and currents took them until running aground or detonating against a suitably hard object. The mines were indifferent to the nature of that hard object, be it a rock, a German warship, or a Russian vessel.

In 1916, the navy general staff established an advanced or outer position at the mouth of the Gulf of Finland, anchored on long-range coastal artillery batteries located on Dagö Island and Odensholm (Osmussaar) Island in Estonia at the south side of the mouth of the Gulf of Finland and Hango and Abo in extreme Southwest Finland on the north side of the Gulf (Achkasov et al. 1979). They laid over 2000 mines across the mouth of the Gulf and established a base at Kuivast on the east side of Moon Island. Further reinforcements to the region came with the assignment of an artillery division from Reval with six batteries of 6 inch (152 mm) guns, and the Russians emplaced long-range, 12 inch guns at Zerel on the Sworbe Peninsula on Ösel Island in 1916. Four additional 12 inch guns went in on Cape Takhona on Dagö Island at the same time (Achkasov et al. 1979). When the ice melted in the spring of 1916, mine layers laid 6000 mines in the Irbe Strait. In late 1916, the navy designated the Baltic Island archipelago a "fortified position" (Melkonov 2003) (Fig. 7.7).

All in all, by early 1917 the Russians had established a matrix of positions in the Gulf of Finland characteristic of Russian land defenses. Three layers of minefields (34,846 mines in all) covered by coastal batteries (284 guns, ranging from 4 to 12 inches in size) formed a formidable barrier behind which their ships could operate (Kozlov 2005), offsetting the German numerical advantage. They also began dredging the Moon Sound between Moon Island and the Estonian mainland to enable their larger ships to use the passage. By 1917, the Sound was dredged to a depth of 9 meters, and the predreadnoughts *Slava* and *Tsesarevich* could use the Sound, avoiding the hazardous night passage through the Irbe Straits under German fire from the Courland Peninsula (Nekrasov 2004).

Summarizing, the Russian naval outer defense position lays at the mouth of the Gulf of Finland. A minefield ran from just north of Dagö Island to Hango in Finland (Fig. 7.8).

Long-range coastal artillery (12 inch) at Toffri on the north cape of Dagö Island facing north and Abo Island on the north side of the Gulf facing to the south protected the direct approaches to the Gulf of Finland from the Baltic Sea. Entrance

Fig. 7.7 Outer defense position

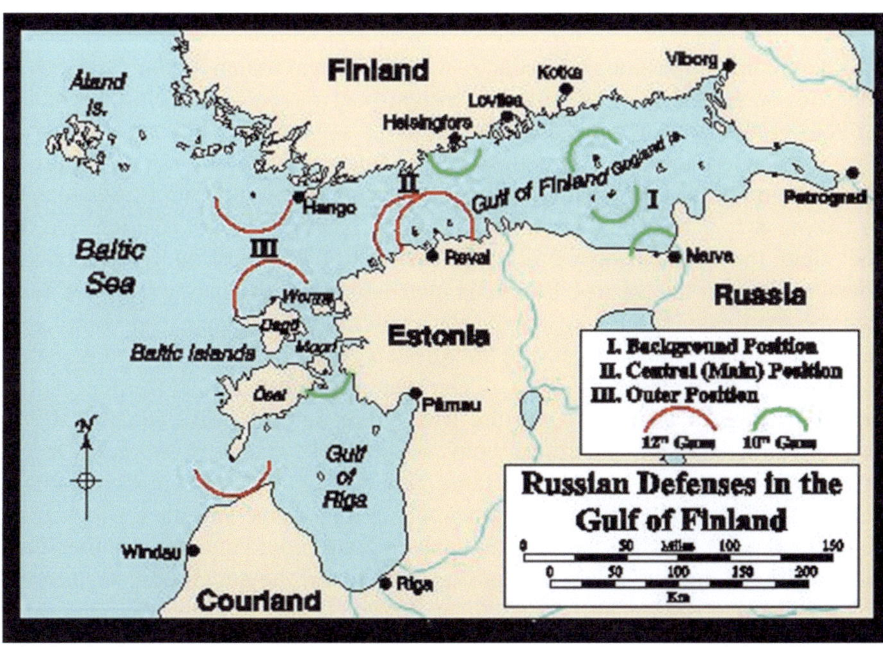

Fig. 7.8 Complete Baltic defenses

into the Gulf of Finland from the south through the Gulf of Riga likewise was blocked by mines and artillery. A fort at Zerel on the southwest tip (Sworbe Peninsula) of Ösel with 12 inch guns covered the formidable minefields of the Irbe Straits on the west entrance to the Gulf of Riga. The Moon Sound, which ran along the eastern side of the Gulf of Riga between the Estonian coast and the Baltic Islands, was narrow, demanded tricky navigation, and was covered by Russian land batteries ranging from 120 to 250 mm along with mines. The dredging of the Moon Sound in 1916 allowed the Russians an escape should the Germans block the western entrance to the Gulf at the Irbe Straits. Unfortunately, the outbreak of the Revolution in 1917 halted progress on the dredging of the sound at a 9 meter depth. Battleships of the newer *Sebastopol* class could not pass (Mawdsley 1978; Nekrasov 2004; Achkasov et al. 1979).

The central defensive position ran across the Gulf of Finland from Helsingfors to Reval. Minefields and coastal batteries located on Nargen and Vulf Islands near Reval and Sveaborg Islands on the Finland side constituted the central position. The Baltic Fleet kept its heavy ships in Helsingfors. Smaller ships used in the Gulf of Riga routinely rotated between Reval and Kuivast. Fighting a qualitatively and quantitatively superior force from behind defensive positions made sense; it also made for an inherently conservative defense that took no risks. The Russians rarely left the Gulf of Finland and for the most part surrendered control of the Baltic to the Germans (Mawdsley 1978). To cite naval historian Paul Halpern, the Russians adopted "...a form of naval trench warfare, with the ships sheltering behind the minefields and coordinating their fire with the powerful coastal batteries" (Halpern 1994).

The remaining question is, of course, was this "naval trench warfare" effective? It depends on the definition of effective. With respect to securing the Gulf of Finland and Petrograd, the answer must be yes, but in the same breath, one must acknowledge that the Germans never actually threatened the region until virtually at the end of the war in the East. In October 1917, a large amphibious operation, *Albion*, saw the Germans rout the Russians from the Baltic Islands defense position. On the other hand, the Germans only broke into the Gulf of Riga past the minefields and coastal artillery at the Sworbe Peninsula after taking the Russian garrison at Zerel from the land side. It must be noted that revolutionary agitation had completely undermined morale and discipline in the Russian garrison. Before their capitulation at Zerel, the Russian big guns drove off German minesweepers, and German capital ships did not even attempt to run the Irbe Straits into the Gulf (Barrett 2008). Perhaps a better assessment of the effectiveness of the "trenches in the ocean" stems from a 1915 episode in the same waters. That year, the Russian predreadnought *Slava* held off several German capital ships in a long-range gun duel. Blocked by the minefields off the Sworbe Peninsula, the Germans could not close on the *Slava* and bring their superior numbers to advantage. Behind the minefields, steaming in circles with relative impunity, the *Slava* drove off the superior German ships (Nekrasov 2004).

What this isolated episode and Operation *Albion* show is that the Russian "trenches in the ocean" would have certainly given the requisite time for the army to assemble sufficient forces near Petrograd in case of an enemy landing. The events also illustrate that if the attacker (in this case the Germans) had the resources and was willing to pay the butcher bill, the defenses could be breached. "Could be," however, is not the same as "were," and Russia's "trenches in the ocean" served their purpose well.

Works Cited

Achkasov, I., Kozlov, I., & Solov'ev, I. (1979). Operations of the Russian fleet in the Baltic Sea, 1914–1917. In N. B. Pavlovich (Ed.), *The fleet in the first world war. Operations of the Russian fleet*. New Delhi: Amerind.

Barrett, M. (2008). *Operation Albion. The German conquest of the Baltic islands*. Bloomington: Indiana University Press.

Cowie, J. (1949). *Mines, minelayers and minelaying*. Oxford: Oxford University Press.

Halpern, P. (1994). *A naval history of world war I*. Annapolis: Naval Institute Press.

Kozlov, D. (2005). Some aspects of operational employment of Russian naval forces in World War I. In *Military thought* 14(2), 208–214.

Luntinen, P. (1997). *The imperial Russian army and navy in Finland 1808–1918*. Helsinki: Finnish Historical Society.

Mawdsley, E. (1978). *The Russian revolution and the Baltic fleet: War and politics, February 1917–April 1918*. London: Macmillan.

Melkonov, J. (2003). *Baterei moonzunda*. RTC Bask: Riga (Latvia).

Menning, B. (1992). *Bayonets before bullets. The Russian imperial army, 1861–1914*. Bloomington: Indiana University Press.

Mitchell, D. (1974). *A history of Russian and soviet sea power*. New York: Macmillan.

Nekrasov, G. (2004). Expendable glory. In *A Russian battleship in the Baltic, 1915–1917*. New York: Columbia University Press.

Pavlovich, N. (1979). *Introduction in the fleet in the first world war. Operations of the Russian fleet*. New Delhi: Amerind.

Podsoblyaev, E. (2002). The Russian naval general staff and the evolution of naval policy 1905–1914 (trans: King, F., & Biggart, J.). *Journal of Modern History*, (66), 42–44, 46–47.

Russian state naval archive (RGAVMF). Fond 417, File 4191. Ship construction program for 1912–1916; Baltic fleet 1911–1915.

Chapter 8
Fort Eben-Emael: Geographic Pivot of the Western Front, 1940

Francis A. Galgano

Introduction

Military geography provides a particularly effective vantage point from which to examine the nexus of terrain and battle; and its study demonstrates, time and again, the central importance of place and that some places are decisive. One such place is a limestone ridge, *Mont St. Peter*, which controls the critical Meuse River crossing sites along the Dutch–Belgian border near Maastricht (Fig. 8.1). So important is *Mont St. Peter* that it has served as a natural defensive position for various armies dating back to the fourth-century Romans (Saunders 2005). It is here that the Belgians built Fort Eben-Emael during the interwar period to exploit the defensive qualities of the natural terrain and the newly constructed Albert Canal. Carefully sited on this 80-meter high ridge of hard limestone, the fort was thought to be all but impregnable. Thus, Fort Eben-Emael was to be the centerpiece of Belgium's defensive system because it provided unrivaled observation across Southern Holland into Germany and because it could control the key entry points into the traditional invasion route between Germany and France.

The decisive terrain at *Mont St. Peter* and the so-called Maastricht Gateway were recognized early on by the Germans, and it historically played a central role in their military planning. Because it is key terrain, it was designated the *schwerpunkt* (i.e., heavy point) for the Schlieffen Plan in August 1914, and von Kluck's 1st German Army made its decisive crossing of the Meuse River here (Possony 1944). Thus, recognizing the vital nature of *Mont St. Peter* to blocking the traditional German invasion route, the Belgian Army constructed the fort at the juncture of the Meuse River and Albert Canal to prevent a similar attack in the future. Fort Eben-Emael, named after a local village (Fig. 8.1), was said at the time to be the most powerful ever constructed and cost nearly 50 million Belgian francs at the time of its

F. A. Galgano (✉)
Department of Geography and the Environment, Villanova University, Villanova, PA, USA
e-mail: francis.galgano@villanova.edu

P. L. Guth (ed.), *Military Geoscience*, Advances in Military Geosciences,
https://doi.org/10.1007/978-3-030-32173-4_8

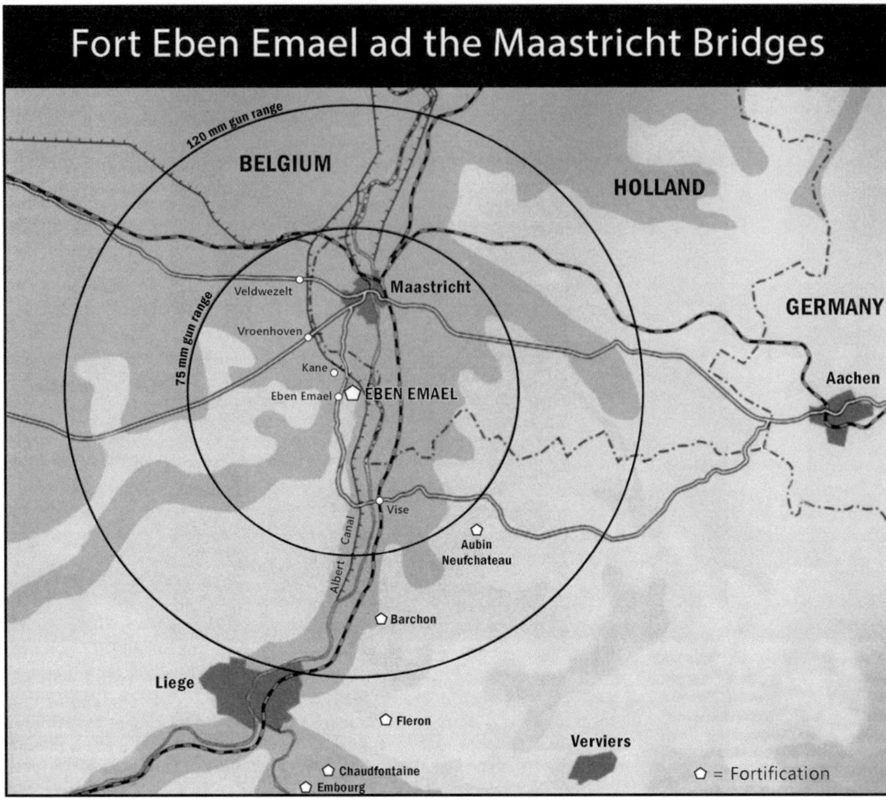

Fig. 8.1 The position of Fort Eben-Emael at the junction of the Albert Canal and Meuse River. The fort's main guns were sited to cover three critical bridges at Kanne, Vroenhoven, and Veldwezelt. (Map adapted from Saunders (2005))

completion in 1936 (Mrazek 1970). When it was completed, it was expected to bar the way into the heart of Belgium, which historically provided invading armies nearly unimpeded access from the east into northern France. Thus, Fort Eben-Emael was to be the lynchpin of Belgium's defensive system and the Allied battle plan. The fall of Eben-Emael on 11 May 1940 was decisive because it opened a lethal gap, which the German panzers violently exploited in an operational feint, making possible their attack through the Ardennes in May 1940.

The Geographic Pivot

In May 1940, the German Army outflanked the Maginot Line, to the south, and main French and British armies in the north, by surging though the Ardennes. The initial German Case Yellow plan (Fig. 8.2) called for an attack that required the

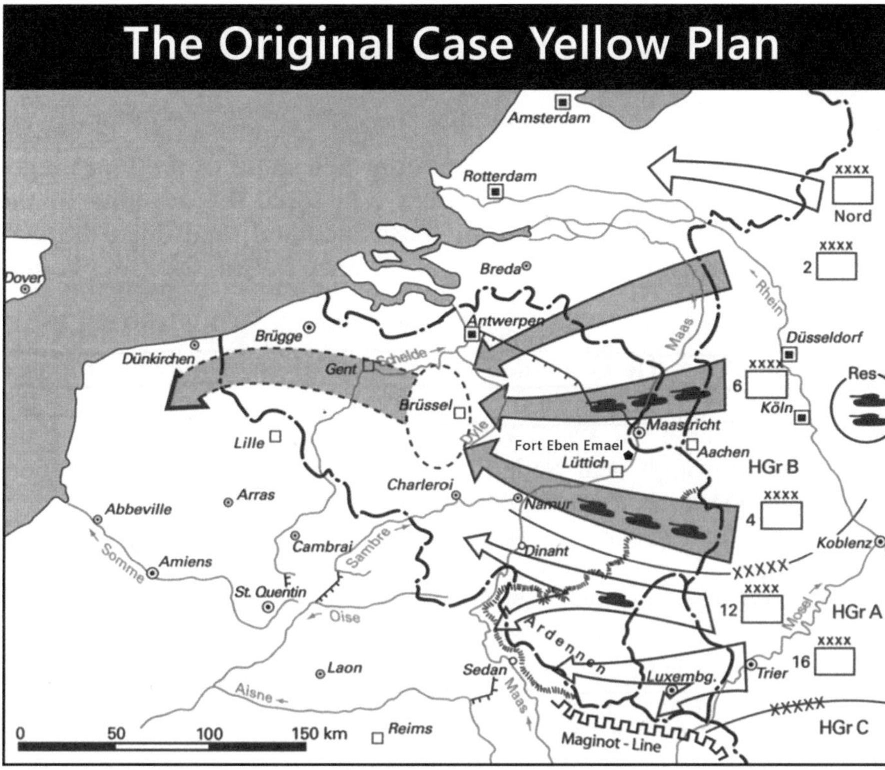

Fig. 8.2 The original Case Yellow plan called for the main attack to fall in Belgium by Army Group B. The schwerpunkt was to be located at Fort Eben-Emael. (Map adapted from USMA (1995))

capture of the fort so that they could gain access into the heart of Belgian territory and make their main effort through Gembloux Gap (Fig. 8.3), which was the traditional invasion route into France. However, Germany eventually adopted von Manstein's more daring plan, which shifted the *schwerpunkt* from Fort Eben-Emael, south to Sedan (Saunders 2005). Nevertheless, the fort was still the key to the new German operational plan, because its capture was designed to draw the main Allied armies deeply into Belgium so that they could be encircled and destroyed. The altered Case Yellow plan exploited the Allies' expectation that the Germans would attack through the Maastricht Gateway much as they did during the World War I. Consequently, von Manstein's variant of Case Yellow—the so-called Sichelschnitt—depended on the use of a "matador's cape" to draw the French and British armies into a trap. Thus, the German attack on Fort Eben-Emael was essential because it served as an operational feint that would widen the frontage of the German offensive, obscure the their main effort in the Ardennes, and fix in place the best formations in the British and French armies, while their rear remained open to exploitation by a panzer phalanx through the Ardennes (Witzig 1973).

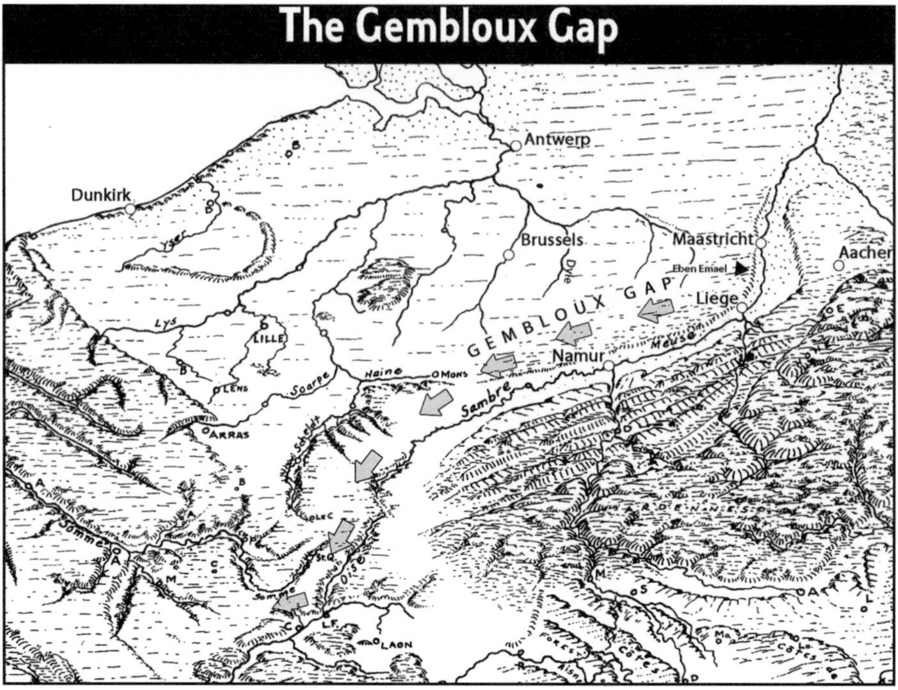

Fig. 8.3 Map illustrating the location of the Gembloux Gap, a drainage divide that offers a high-speed invasion route between Franca and German. (Map adapted from MIS (1943))

The Loss of Eben-Emael: Prelude to Military Disaster

Fort Eben-Emael, with its heavy steel turrets and massive concrete casemates, was thought to be invulnerable. It was built to serve as the lynchpin of the fortified zone known as the Albert Line. As such, the fort was expected to hold out for many months against a traditional infantry assault, although the French and British were looking only for 6 days so that they could establish a coherent defense along the Dyle River in Belgium. Though the Belgian garrison was ten times the size of the German assault force and its guns were protected by tons of reinforced concrete and steel, the fort fell after only 28 hours of fighting (Mrazek 1970).

The loss of the fort was decisive because two German panzer divisions surged through this fatal gap and engaged French mobile forces in Belgium, while Guderian's panzers penetrated the Ardennes to the south and encircled them from the rear. The Allies were stunned and unnerved by the mysterious and unorthodox attack that captured the fort and the Belgian garrison of nearly 700 soldiers surrendered after little more than a day (Axelsson 1940, Witkowski 2004). The world was shocked by the audacity of the German plan, and reports of secret magnetic weapons, death rays, and nerve gas appeared in newspapers (New York Times 1940).

Clearly, the loss of the fort fixated the Allies and all but confirmed that the Germans intended to exploit the Maastricht Gateway as they did during the World War I, and the best Allied units raced headlong into Belgium to meet the advancing Germans. Thus, they fell into a deadly trap as the main German panzer attack penetrated to their south through the Ardennes. When Guderian's panzers emerged at Sedan, French forces along the Ardennes collapsed in 13 May, ultimately precipitating the Allied debacle at Dunkirk (von Manstein 1958).

So, how did this military disaster come about? Although Fort Eben-Emael occupied a strategic geographic position and was seemingly impregnable, it was manned by an inadequate and poorly trained garrison. Furthermore, the fort had intrinsic flaws—a lack of infantry positions, balky equipment, and inadequate defenses on its flat superstructure that hastened its demise (Kaufman and Jurga 1999; Saunders 2005). The German plan magnified and clearly exploited the fort's weaknesses. The Germans also employed emerging technologies to support their audacious operation and thus were able to surprise the Belgians who never anticipated an assault directly on the fort's lightly defended superstructure. This battle witnessed the first operational use of the hollow charge and assault glider (Witzig 1973). The hollow charge gave the relatively small and lightly armed paratroop assault force a weapon capable of defeating the fort's massive armored cupolas and concrete casemates. The assault gliders permitted the Germans to approach the fort silently and land within meters of key fortifications (Mrazek 1970; Witzig 1973). Thus, the great fort fell almost before the battle in the West began, and with it, the Allies lost an essential bulwark of their strategic plan (Bassett 1948; Guderian 1957; Manstein 1958).

Background

The main German attack through the Maastricht Gateway during the opening days of WWI demonstrated to the Belgians the critical nature of the terrain at *Mont St. Peter.* Their failure to defend this site adequately nearly cost them the whole of their country and the Allies the war. As the Nazi Party took hold in Germany during the interwar period, it became readily apparent to Belgian leadership that once more their country would likely serve as the invasion route that a resurgent German Army would use to attack France. Hence, they undertook a comprehensive review of the nation's defenses (Dunstan 2005; Horne 1972).

Despite their realization that their country stood in the way of the Germans, their options were limited because World War I shattered the Belgian economy, and they were in no position to obligate an enormous budget to a significant upgrade of their ground forces. However, they were impressed by the resilience of the French forts at Verdun and encouraged by the victorious Allies' general belief in the power of defensive systems. Hence, the Belgians decided to face the German threat with a strategy that relied on powerful fortifications to block and frustrate a German attack, while Allied armies could rally to their defense (Saunders 2005). The new system of fortifications included refurbished structures at Liege and Namur but would also take advantage of the Meuse River and newly constructed Albert Canal (Figs. 8.1 and 8.3). These natural barriers and fortifications provided defensive depth and covered the vital river crossing sites in the Maastricht Gateway. By relying on a system of forts, the Belgians were hoping to economize, provide time for their army to mobilize, and solve their fundamental problem: a geographic situation that made the country a natural corridor for invading armies.

Case Yellow

The German operational plan for the invasion of France was called Case Yellow and evolved through two primary iterations during the fall and winter of 1939–1940 (Manstein 1958; Guderian 1957). In its earliest form, the key element of the plan was the use of Belgium's favorable terrain for the main attack so that German panzer forces could outflank the Maginot Line by entering the Maastricht Gateway and exploiting the Gembloux Gap with Army Group B (Horne 1969). Throughout the plan's evolution, which ultimately changed Army Group B's role to that of an operational-level feint, the Germans were forced to consider how to overcome Fort Eben-Emael, which was the lynchpin of the Allied defensive system (Bassett 1948; Horne 1972).

The strength of Fort Eben-Emael was troublesome for the German high command. Once it became clear that the French and British were not prepared to accept a negotiated peace, the likelihood that the German Army would have to come to grips with the fort was magnified. In the original version of Case Yellow (Fig. 8.2), the German General Staff developed a plan that was based loosely on the *Schlieffen Plan*, with the main attack essentially replicating von Kluck's first war attack through the Maastricht Gateway (Dunstan 2005). The Germans planned to deploy a main striking force (i.e., Army Group B) of 37 infantry and 6 panzer divisions to exploit Belgium's excellent terrain. This plan also called for a supporting attack by Army Group A of some 27 divisions through the Ardennes (Horne 1969). To make this plan work, the Germans knew they would have to neutralize Fort Eben-Emael and secure the bridges over the Albert Canal opposite Maastricht very quickly. Accordingly, *Sturmabteilung Koch*, a highly trained force of paratroopers, was formed to make this daring attack on the fort and the adjacent bridges. These highly trained glider assault troops began to develop the tactics needed to quickly neutralize the great fort.

Because of the Western Front's strategic geometry (Galgano 2008), and strength of the Maginot Line, the Allies were expecting the Germans to attack through Belgium as they did during the WWI (Horne 1969; Dunstan 2005). For that reason, the French intended to engage the Germans in Belgian territory once the Germans violated its neutrality. This operational concept was driven by France's fundamental strategy to avoid fighting in its northeastern industrial region around the city of Lille (Jackson 2003). However, for the French plan to engage the Germans forward in Belgium to succeed, the Albert Line fortifications, and Eben-Emael specifically, had to delay the Germans for 5 or 6 days so that the British and French armies could occupy their positions along the Dyle River and block the Gembloux Gap. The Belgian Army was expected to withdraw from the border to defend along the Albert Line. This entire plan depended on a gigantic wheeling motion of Allied forces into Belgium, and Fort Eben-Emael was the vital pivot for this maneuver (Mrazek 1970).

Following the fall of Poland, the Germans initially intended to attempt their invasion of France in the late fall of 1939. However, after a number of weather-related delays, the Germans were finally prepared to conduct Case Yellow on 17 January 1940 (Manstein 1958). As fate would have it, however, a German plane carrying a Luftwaffe staff officer was forced to land in Belgium 1 week before the invasion because of bad weather. The staff officer was carrying a major component of the Case Yellow plan, which was captured by the Belgians. Unluckily for the Allies, the captured plan confirmed that the Germans intended to make the main assault through the Gembloux Gap, much as they expected. The Allies went on full alert and their subsequent movements, which were observed by the Germans, demonstrated their operational intentions, and they confirmed that Sedan was to be weakly defended and that the main Allied armies could be outflanked and surrounded using an attack through the Ardennes (Dunstan 2005).

These events convinced Hitler and the reluctant General Staff to accept von Manstein's daring operational concept, which fundamentally altered Case Yellow. His plan changed the main assault to Army Group A by allotting it 44 infantry divisions and 7 of Germany's 10 panzer divisions to the attack through the Ardennes. More importantly, von Manstein's plan placed the *schwerpunkt* at Sedan, which was defended by third-rate French reserve divisions (Manstein 1958). This was an audacious plan that depended on an unprecedented concentration of armor to cross the Ardennes and breach the Allied line at Sedan. Although the plan was risky, if successful, it would deliver a decisive blow to the Allies and eliminate their qualitative and quantitative superiority on the battlefield (Horne 1969; Barry 1972; Jackson 2003).

Regardless of this fundamental shift in German strategy, von Manstein's plan depended upon the ability of Army Group B to make an operational-level feint and draw the Allied armies into Belgium so that they could be surrounded and destroyed (Dunstan 2005). What the Germans needed was a "matador's cape" to entice the Allies into the trap, and Fort Eben-Emael was going to serve that purpose. Thus, it maintained its role as the geographic pivot of the Western Front. Although Army Group B would no longer make the main attack, *Sturmabteilung Koch's* mission to reduce Fort Eben-Emael was essential because the ultimate success of Case Yellow depended on their ability to focus Allied attention on the key terrain in the Maastricht Gateway and divert their attention from the movement of the main German effort through the Ardennes along their southern flank (Mrazek 1970).

Belgium's Geographic Predicament

It is said that you can make your own history, but you have to live with your geography. Sadly, the Belgians have been forced to live with their historic geographic dilemma, which made their country into a battleground. The Belgian terrain is a perfect corridor for military forces by connecting France and Germany—the seminal European rivals during the past two centuries—and it includes some of the best terrain in Northern Europe for mobile warfare. The critical geographic feature is the Gembloux Gap, a drainage divide between the Dyle and Meuse Rivers (Fig. 8.3). The Gembloux Gap permits an army to traverse central Belgium between Germany and France without having to conduct a river crossing. Consequently, in a military sense, Belgium serves as a conduit for invading armies. Belgium's geographic predicament is heightened because some of the best roads and railways in Europe dominate its landscape as well (Mrazek 1970).

This unique geography caused Belgium to live in fear of being trampled by its much stronger and more warlike neighbors. This trepidation was assuaged occasionally by international agreements intended to designate Belgium as a neutral buffer during times of war. Nonetheless, Belgium's position was always rather precarious and it seldom felt secure (Mrazek 1970). This problem was made worse as Germany and France fortified their common border, making a flanking march through Belgium a more attractive alternative to a bloody frontal attack. As the 1800s drew to a close, it became increasingly likely that Germany or France would use Belgium as a military corridor notwithstanding international agreements. Consequently, Belgium began a program of fortifications in the Meuse River valley in an attempt deny its favorable terrain to a potential aggressor, and Fort Eben-Emael was the pinnacle of these efforts (Saunders 2005).

The Evolution of Fort Eben-Emael

Belgium's unfortunate military geography necessitated the need for the development and implementation of a comprehensive system of fortifications to discourage potential aggressors from violating its neutrality. By the late 1800s, it became increasingly apparent to the Belgians that both France and Germany had designs on exploiting its excellent terrain to avoid costly battles along the Franco–German border. Consequently, they developed fortified zones around their principle cities, such as Liege, Namur, and Antwerp (Saunders 2005).

By 1914, the rings of fortifications built around Belgium's key cities were essentially complete, but they suffered from several glaring deficiencies. Perhaps the single most important flaw in Belgium's defensive system was the absence of a credible fort to cover the vital crossing points in the Maastricht Gateway. The fort recommended for this region was never constructed because of a lack of funding. General Brialmont, who developed the Belgian fortification system, lamented this situation and declared that "[Belgium] will weep tears of blood for not having built that fort" (Saunders 2005, p. 11). His words were prophetic because in 1914, the Germans were presented with a clear and open route into the heart of Belgium.

In August 1914, the Belgian forts failed too because they were not designed to withstand modern artillery. The Belgian's inexplicably used a poor grade of concrete and did not reinforce many of their structures to withstand modern artillery (Mrazek 1970). Thus, German siege artillery was able to quickly reduce the forts. Nevertheless, after the Great War, Belgium remained committed to developing large, static fortifications to protect its territory. The Belgian's postwar analysis convinced them that if a fort had been sited on *Mont St. Peter*, it might have delayed the German crossing of the Meuse River enough to disrupt their timetable and the outcome of the war may have been different (Mrazek 1970).

Genesis of the New Albert Line and Fort Eben-Emael

Between the wars, the Belgians were convinced that powerful defenses would have to be built in order to deter a resurgent Germany from violating its territory to outflank France's Maginot Line. Thus, in 1930, a special military commission recommended building a new defensive line beyond the Meuse River and along the newly constructed Albert Canal (Mrazek 1970). This plan was formalized in 1931 when the government approved a new national defense strategy that incorporated four principle components:

1. The improvement and reconstruction of the ring of pre-WWI forts surrounding Liege. These were called PFL (*Position Fortifiée de Liège*) II (Fig. 8.1).
2. The construction of bunkers and fortifications to protect the line of the Albert Canal, to included pre-chambered bridges.
3. A new line of modern forts along the border with Germany, south of Maastricht (this line was called PLF I).
4. The lynchpin of the entire system was the construction of a large, powerful fort on *Mont St. Peter* to defend the Maastricht Gateway. This fort was critical too, because it linked PLF I to the fortifications of the Albert Canal and PFL II. Thus, Fort Eben-Emael was the lynchpin of the Belgian defensive system.

Geography of Fort Eben-Emael

When Fort Eben-Emael was completed in 1935, it was quite literally built into the hard limestone of *Mont St. Peter*, which was an imposing natural feature in its own right. The hill mass rose some 80 meters, with near vertical cliffs above the Meuse River. The site was made more imposing after the completion of the Albert Canal, which connected the Meuse River 3 miles south of Maastricht to Belgium's northern

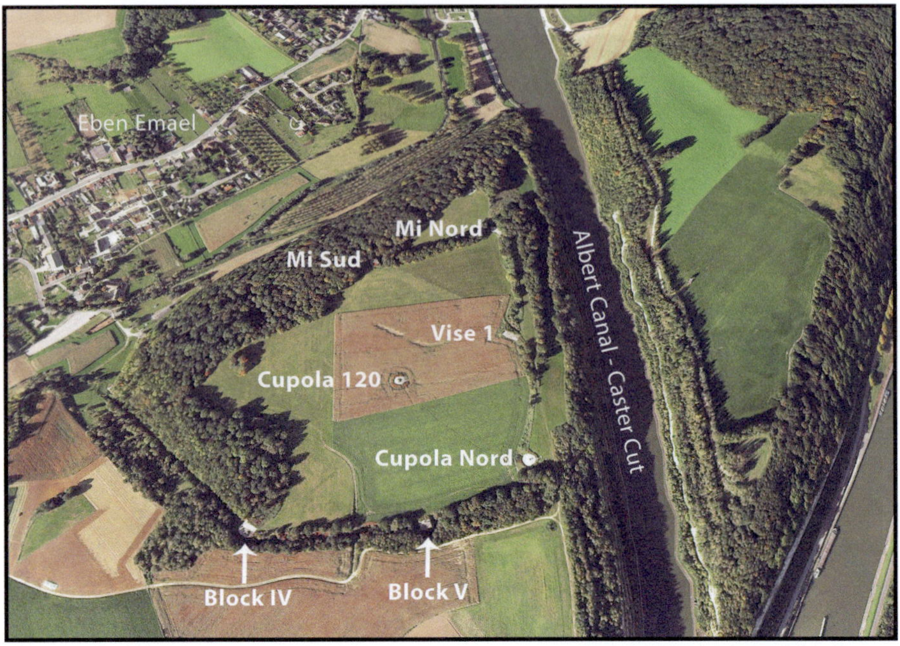

Fig. 8.4 Image of Fort Eben-Emael. Several of the major components of the fort are visible and labeled as is the Albert Canal–Caster Cut. (Google Earth (accessed 4 December 2017))

ports. To complete the canal, Belgian engineers were required to cut through the ridge's limestone—an impressive engineering feat that created the Caster Cut. The cut is a trench hewn through the northeastern flank of the mountain with near vertical walls nearly 80 meters high. Thus, the Caster Cut and canal provided excellent protection to the fort especially its eastern flank (Fig. 8.4). Furthermore, *Mont St. Peter* afforded an incomparable view of the countryside into Germany, only 24 km to the east (Mrazek 1970).

The fort occupies (it is still there today) a roughly triangular area that measures 830 meters long by some 730 meters wide at its southern base. The fort was tunneled directly into the mountain in three levels with only its casemates, observation cupolas, and gun turrets exposed. The Albert Canal and Caster Cut protect the fort's flanks in the northeast, and a wet moat and numerous obstacles steepen the northwest approaches (Figs. 8.4 and 8.5). Additionally, the small stream along this flank was engineered to flood the valley, making a landward approach in this direction virtually impossible. The southern slopes are defended by 4-meter reinforced concrete walls, a moat, and numerous antitank obstacles. Consequently, Fort Eben-Emael was well protected against a conventional infantry ground assault (Saunders 2005).

The fort's designers hoped that Dutch forces would slow any German advance across the so-called Maastricht Appendage and thus provide much needed time for Belgian forces to properly occupy the fort and destroy the Meuse River bridges at

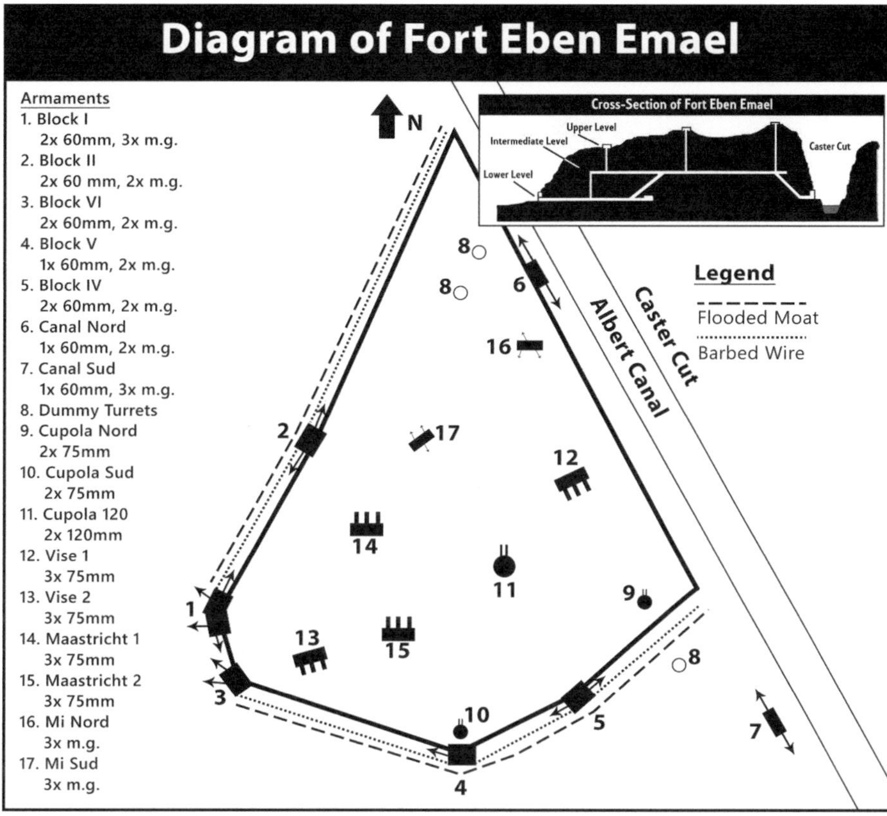

Fig. 8.5 Diagram illustrating the major components of Fort Eben-Emael. (Adapted from Kaufman and Jurga (1999))

Maastricht (Fig. 8.1). The newly completed Albert Canal included three important bridges north of Fort Eben-Emael at Veldwezelt (7 km north), Vroenhoven (3.2 km north), and Kanne (1.6 km north) (Fig. 8.1). Each bridge was protected by a dedicated infantry unit, bunkers, and pre-chambered demolition charges. The fort's guns were sited to provide artillery support to infantry units defending the bridges, and should the explosive charges fail to collapse the bridges, the fort's guns had the mission to destroy them (Kemp 2006).

Notwithstanding its size, there was little evidence of Fort Eben-Emael on the surface because it was built by excavating directly into *Mont St. Peter*. One entered the fort at Bloc 1 (Fig. 8.5), some 60 meters below its superstructure. The fort included accommodations for 1400 soldiers and included generators, air filtration units, water purification units, magazines, mess facilities, and barracks; and it was expected to hold out for several months. The perimeter was defended by five large blockhouses along with moats and extensive obstacles. The superstructure boasted 17 observation cupolas, 4 artillery casemates, 3 armored turrets, and 2 machine-gun casemates (Kemp 2006).

The fort's firepower was divided into two discrete groups: an offensive (i.e., *1st Batterie*) and a defensive battery (i.e., *2nd Batterie).* The centerpiece of the fort's offensive battery was the massive *Cupola 120* (Fig. 8.5), which was situated in approximately the center of the superstructure. This was a large, fully rotating, armored turret mounting 120 mm guns capable of engaging targets 18 km distant. The fort was limited to this gun caliber because the Belgians did not want to mount guns that could reach Germany and perhaps jeopardize their neutral status. In the end, however, this shortcoming did not cause the fort to fall (Saunders 2005). Two other armored turrets were located on the fort's superstructure, *Cupolas Nord* and *Sud*, each mounting twin 75 mm guns capable of striking targets 11 km distant. Each turret could rotate through 360° and engage targets in any direction. Only *Cupolas Nord* and *Sud* could retract below the level of the superstructure (Kemp 2006).

The fort's offensive battery also included four large casemates. Two were oriented north to cover the crossing sites at Maastricht (i.e., *Maastricht 1 & 2*) and two oriented to the south to cover the crossing sites in the Vise Gap (i.e., *Vise 1 & 2*) (Fig. 8.5). These were massive concrete structures, and each casemate mounted three 75 mm guns that could fire through a 70-degree arc (Kemp 2006). Thus, the fort was somewhat under-gunned considering its full potential. Nevertheless, the guns could easily carry out the intended mission of the fort, which was to provide fire support for the infantry defending vital crossing sites in the Maastricht Gateway and place uninterrupted fire on German forces as they advanced into Belgium (Witzig 1973; Mrazek 1970).

The *2nd Batterie* comprised Fort Eben-Emael's defensive capability, which was formidable against a traditional ground assault. These defenses included five large bunkers located at angles along the base of the fort, armed with search lights, heavy machine guns, and 60 mm antitank guns (i.e., *Blocs I–VI* (note that there is no *Bloc III*); Fig. 8.5), along with two large bunkers (i.e., *Canal Nord* and *Canal Sud*) sited at the level of the Albert Canal in the Caster Cut. The bunkers covered the approaches to the fort, which were also guarded by barbed wire entanglements, moats, and antitank ditches. The superstructure was defended by two machine-gun bunkers (i.e., *Mi Nord* and *Mi Sud*; Fig. 8.5) along with a battery of four antiaircraft machine guns. The defensive battery also included *Bloc 01*, which is situated outside of the fort's footprint to directly observe the Lanaye Locks and monitor the countryside into Germany. Lastly, the fort's defenses included three dummy 120 mm cupolas designed to confuse enemy aerial spotters and draw artillery fire away from actual positions. In practice, the dummy positions turned out to be quite effective because the Germans failed to ascertain their true nature and devoted a team of paratroopers to neutralize the northern two dummy turrets (Witzig 1973).

Fort Eben-Emael was perhaps more powerful than any single fort in the Maginot Line. However, for all its strength, the fort manifested important shortcomings. In 1940 the fort was devoid of trees except along its northern perimeter, and the superstructure was essentially a large open field with few obstacles—clearly the Belgians did not foresee the potential for an airborne or glider-borne assault (Witzig 1973). The fort did not include a detachment of interval troops to fight outside of the structure, and certainly, there were no trenches or other positions from which the garrison could defend against intruders on the superstructure. Unlike the Maginot Line, the fort did not have an internal transportation system, so the garrison was forced to walk very long distances. This was decisive because the glider assault, which began just as the Germans crossed into Holland some 24 km distant, meant that many of the positions were empty when the gliders landed (Witzig 1973). Inexplicably, many of the fort's critical systems, such as the ammunition hoists, did not work properly at the supreme moment, and soldiers were forced to carry artillery ammunition from the deeply placed magazines to the turrets on the superstructure. This problem was exacerbated because the ready ammunition in the turrets and casemates did not include canister rounds. These could have perhaps tipped the scales during the decisive moments early in the attack when the German paratroopers were just beginning to reduce the fort's superstructure (Dunstan 2005). Finally, the cornerstone of the Belgian defensive system was simply manned by a poorly trained garrison of unmotivated soldiers. Duty in the fortifications provided little opportunity for promotion in the Belgian officer corps, and the soldiers that operated the forts were often of a lesser quality than those in infantry and cavalry units. Thus, Fort Eben-Emael, for all of its strength and critical position as the pivot of the Western Front, was perhaps doomed to failure (Manstein 1958; Witzig 1973; Kemp 2006).

Glider Assault: 10 May 1940

The Germans realized that the success of their attack in the West would hinge on a rapid crossing of the Meuse River/Albert Canal and exploitation of Belgium's favorable terrain. Time was critical for the Germans because the French and British held a clear qualitative and quantitative superiority in artillery and armor. The principle German advantage lay in their superior training, doctrine, and leadership (Barry 1972). A protracted siege of Fort Eben-Emael would erase their slim advantages and play into Allied hands. Consequently, the Germans developed a daring plan that employed two emerging technologies—the assault glider and hollow charge explosive—to reduce the fort quickly and ensure a rapid exploitation of the Gembloux Gap and employment of their *Blitzkrieg* doctrine.

The centerpiece of the plan was a unit of highly trained paratroopers called *Sturmabteilung Koch*, named after its commanding officer. Although the Germans originally considered using paratroopers in a conventional role, it was realized that too few would land on the fort despite its large size. More importantly, even if they landed in adequate numbers, it was doubtful if the lightly armed paratroopers could destroy the armored turrets and heavy casemates (Witzig 1973). Hence, they decided to use a relatively new weapon, the DFS 230 assault glider, which was capable of pinpoint landings on the fort's flat roof. Each glider could carry 11 paratroopers and the heavy demolition charges. Furthermore, the gliders could be released from their tugs on the German side of the frontier and make a silent descent onto the fort's surface at dawn – at the precise moment German forces crossed into Dutch territory. This was perhaps the critical advantage because the Belgians needed time to fully alert and occupy the fort – time that was expected to be provided by Dutch resistance to the initial German ground attack. The glider landing achieved complete surprise, and many of the fort's vital defenses were not operational as the paratroopers charged out of their gliders and began their systematic and highly coordinated reduction of the fort's critical structures (Mrazek 1970).

The second technological surprise used by the Germans was the hollow charge, which was powerful enough to penetrate the fort's armored turrets, observation cupolas, and concrete structures (Mrazek 1970). In practice, the hollow charges were decisive because they devastated many of the fort's supposedly indestructible structures. Preliminary reports received by the fort's commander on the morning of the attack indicated only that there were small numbers of enemy soldiers on the superstructure and then, to his astonishment, the key structures were being smashed by mysterious explosions. The hollow charges blew holes into the armored cupolas, vaporizing their crews, and pulverized the concrete casemates (Dunstan 2005). Within minutes, many of the fort's offensive and defensive batteries were out of commission, and German paratroops gained access to the upper levels of the superstructure (Witzig 1973). The Belgian garrison, which was comprised of second-rate soldiers, was quickly demoralized and surrendered some 28 hours later (Axelsson 1940).

The sub-unit of *Sturmabteilung Koch* designated to land on the fort was called *Sturmgruppe Granit* and was commanded by Lieutenant Witzig. *Granit* consisted of 11 squads (78 soldiers) of highly trained infantry and assault engineers. The force trained in complete secrecy for 5 grueling months. Each squad could accomplish the tasks of any other. Their level of training manifested itself at the very start of the operation when the commanding officer's glider was inadvertently released too soon and he failed to land in the initial assault. Despite this setback, *Granit* carried out its attack to near perfection and, by nightfall, controlled the fort's superstructure. Only *Cupola Nord* remained in operation, and its crew was finally forced to disable the turret after a few hours. The world's strongest fort fell to a small group of highly trained assault troops, which lost only six men killed during the daring attack (Witzig 1973).

Summary and Conclusions

Fort Eben-Emael proved to be the geographic pivot of the Western Front. Allied and German plans hinged on the fort and its crucial terrain: the Allies needed the fort to slow the German attack and give them time to deploy, and the Germans needed to destroy the fort to draw the Allies into their *Sichelschnitt* trap. To the Germans, the seizure of the fort was part of an important operational-level feint, intended to confirm the Allied analysis that the main attack would fall in the Gembloux Gap. The fort was the essential lynchpin in the Allied defensive system; nevertheless, it rapidly fell to a small, highly trained force of German glider troops. Clearly, the world was awed by the mysterious and unorthodox attack. The world's strongest fort fell to one of the most audacious special operation raids in history, and the results were decisive. The collapse of Eben-Emael opened a lethal gap in the Allied line, through which German panzers advanced, thus drawing the Allies into a strategic cul-de-sac. Within 3 days, the Germans burst out of the Ardennes at Sedan, ultimately reaching the English Channel, precipitating the Allied debacle at Dunkirk. Ten days after the fall of Eben-Emael, the Belgians capitulated, and 32 days afterward, France surrendered.

Fort Eben-Emael was built during the interwar period to protect the crossing points along the Meuse River and access point into the Gembloux Gap. The fort was carefully sited on the key terrain south of Maastricht and protected the all-important crossing sites over the Meuse River. Fort Eben-Emael was the key installation of the Belgian defensive system because it served as the hinge upon which the Belgian Army maneuvered into their Albert Line defenses. The fort also linked the Albert Canal and Meuse River defenses to those of the PFL I and II forts. Most importantly, however, Fort Eben-Emael controlled the vital Maastricht Gateway.

The fort was equally vital to the Germans because they needed to neutralize its defense in order to cross the Meuse River and Albert Canal in a timely manner and threaten the Gembloux Gap. This was essential for them to expand their attack and demonstrate the "matador's cape," thus drawing British and French into Belgium while at the same time concealing their main effort in the Ardennes Forest to the south. Hence, *Sturmgruppe Granit's* intrepid assault was fundamental to the events that followed on the Western Front because the Germans were able to achieve strategic surprise at Sedan and ultimately surround and eliminate the bulk of the Allied armies in Belgium. Only the miracle at Dunkirk saved the Western Allies from total defeat.

Although Eben-Emael was perhaps the strongest fort ever built, it suffered from several serious flaws. The fort's superstructure lacked sufficient defensive works to defeat an infantry assault. Clearly, the Belgian's did not consider the possibility of a glider-borne assault, and the Germans achieved a decisive tactical surprise. The garrison did not include interval troops that could defend the fort's casemates from infantry. Instead, the Belgians depended on a few heavy machine-gun casemates to sweep the top of the fort. Once *Sturmgruppe Granit* destroyed these—in some cases before they were manned—the collapse of the fort was only a matter of time. Finally, the Belgians invested a national treasure (some 50 million francs) into this important fortification and then staffed it with poorly trained, second-rate officers and soldiers.

In the final analysis, the examination of Fort Eben demonstrates the central role of terrain in warfare. The fort and the sensitive Maastricht Gateway were central elements of the German and Allied plans. In every war, there is usually one place that is clearly decisive: *Mont St. Peter* has been a vital defensive position for more than a millennia, and in 1940, it was perhaps the single most important piece of ground in the war. It was the pivot upon which the entire battle turned.

References

Axelsson, G. (1940). Berlin lists gains. New York Times, 12 May 1940, *1*.

Barry, R. H. (1972). *Military balance. History of the second world war* (pp. 334–356). New York: Marshall Cavendish USA, Ltd.

Bassett, J. A. (1948). Past airborne employment. *Military Affairs, 12*(4), 206–216.

Dunstan, S. (2005). *Fort Eben-Emael: The key to Hitler's victory in the west*. Oxford: Osprey Publishing, Ltd.

Galgano, F. A. (2008). Sichelschnitt: Geographical analysis of a decisive campaign, France 1940. In P. Nathanail, B. Abrahart, & R. Bradshaw (Eds.), *Military geology and geography: History and technology* (pp. 185–198). Nottingham: Total Graphics, Ltd..

Guderian, H. (1957). *Panzer leader*. New York: Random House, Inc..

Horne, A. (1969). *To lose a battle, France 1940*. New York: Penguin Books.

Horne, A. (1972). *Breakthrough at Sedan. History of the second world war* (pp. 356–372). New York: Marshall Cavendish USA, Ltd.

Jackson, J. (2003). *The fall of France: The Nazi invasion of 1940*. Oxford: Oxford University Press.

Kaufman, J. E., & Jurga, R. M. (1999). *Fortress Europe*. Conshohocken: Combined Publishing.

Kemp, I. (2006). *Fort Eben-Emael*. Hersham: Ian Allan Publishing, Ltd..

Military Intelligence Section (MIS). (1943). *Continental Europe strategic maps and tables part I: Western Europe*. Washington, D.C.: Army Map Service.

Mrazek, J. E. (1970). *The fall of Fort Eben-Emael*. Washington, D.C.: Robert B. Luce, Inc..

New York Times. (1940). Nazis report use of secret weapon. 13 May 1940, *2*.

Possony, S. (1944). May 1940: The pattern of bad generalship. *Military Affairs, 8*(1), 33–41.

Saunders, T. (2005). *Fort Eben-Emael*. South Yorkshire: Pen and Sword Books, Ltd..

United States Military Academy (USMA). (1995). *The west point atlas of war: World War II: European theater*. Washington, D.C.: U.S. Defense Printing Service.

von Manstein, E. (1958). *Lost victories*. Novato: Presidio Press.

Witkowski, P. (2004). Glider assault on Eben-Emael as an archetype for the future. *Infantry, 93*(2), 28–34.

Witzig, R. (1973). *Coup from the air: The capture of Fort Eben-Emael* (pp. 386–397). New York: Marshall Cavendish USA, Ltd.

Chapter 9
Early Virtual Reality Simulators for the Fleet Air Arm and Royal Navy

Shane Guy

Introduction

Virtual reality simulators became increasingly common in the 1980s due to developments in computer processing, graphics and projection technology. The well-known Microsoft Flight Simulator was launched in 1981 and now developed and rebranded as Prepar3d (by Lockheed Martin) encompasses a full range of environments and means of transport from underwater to suborbital. However attempts at virtual reality simulators, or 'synthetic trainers' as they were known, started in the 1920s. They included realistic aircraft on moving platforms, and early attempts to incorporate these in an environment had started before the Second World War. The leading simulator at this time was the Link Trainer by the Link Aviation Devices, now part of L3 Communications. These were used extensively by air forces, but at best the Link Trainer was enclosed in a crude screen with a rough landscape painted inside (Corry 1984). Fit Ups, a theatre scenery supplier in Manchester, following a chance meeting when trying to tender for camouflage work at the Admiralty in 1940, became involved in trying to improve the realism of the Link Trainers initially using theatrical scenery techniques. The Fleet Air Arm was keen to develop the use of the Link Trainer for training pilots in torpedo bombing. Fit Ups, the Strand Electric Co. and the JVW Corporation – 'Suppliers of the Link Trainer to the British Empire' (Flight 1939) – were commissioned to develop a synthetic trainer comprising of a simulated aircraft that moved in all three axes, enclosed by a projection screen, or cyclorama, on which would be projected a realistic target and atmospheric effects.

S. Guy (✉)
School of Art and Design, Nottingham Trent University, Nottingham, UK
e-mail: shane.guy@ntu.ac.uk

P. L. Guth (ed.), *Military Geoscience*, Advances in Military Geosciences,
https://doi.org/10.1007/978-3-030-32173-4_9

Early Flight Simulators

The need for safe flight training was identified early in the history of powered flight. The Wright brothers considered it essential and learnt to fly gliders in 1901–1902, prior to their success at Kitty Hawk, North Carolina, with powered flight in 1903. Several of their competitors, including Maxim and Langley, considered that aircraft and engine design were more important than pilot training, with disastrous results when they tested their creations.

Early developments in trainers generally took one of two forms. First, a real aeroplane, with reduced wing surfaces such as the cut-down Bleriot monoplane (Williams 1929). These were incapable of flight but would respond fully on the ground, including taxiing and reacting to the prevailing wind. Alternatively a structure on some sort of gimbal, with controls and control surfaces, such as the trainer by Eardly Billings shown at the Stanley Air Show 1910 (Flight 1910) or the better known Sanders Teacher. These incorporated actual aeroplane components which could be transferred when the pilot was ready to construct their own aircraft (Haward 1910). Trainers grew in sophistication but in similar forms until the 1920s.

Ed Link, the inventor of the Link Trainer, was keen to learn to fly, but the cost in 1927 was very high. Having learnt the controls and to taxi on a friend's plane, he decided to develop a better method (L3 Simulation and Training 2012).

After 18 months of work, the Link Trainer was launched. This consisted of a mocked-up aircraft on a two-axis mount similar to that used for the previous 20 years; however Link achieved a radical improvement in the response of the aircraft. Although the controls continued to move the rudder, ailerons, etc., by means of pneumatic controllers, the plane would roll and pitch fluidly in response to the controls just like an actual aircraft. Enhancements quickly included instrumentation, 360° rotation and stall buffeting, using additional pneumatic bellows. Control surface 'feel' was improved by hydraulic dampers to provide the resistance due to airflow in flight. The simulator controls in the instructor station recorded the path of the aircraft by means of a motorised 'crab', a small robot, moving across a plotting table, with an inked wheel whose speed and direction were tied to the trainer.

The Crail Prototype

The Royal Naval Air Station at Crail in Fife, Scotland, HMS Jackdaw (Bentham 1992), was selected as the site of the first Torpedo Attack Teacher (TAT, Fig. 9.1). This was the first attempt to place a simulated aircraft within an interactive battle environment. This consisted of a modified Link Trainer enclosed in a 300° screen, full colour floodlighting of the screen, weather and lighting effects, with a projected target that moved realistically on the screen. The instructor's desk also included the controls for speed and heading of a target ship, which was converted to a location by a second crab recorder on the plotting table.

Fig. 9.1 The Crail Prototype. (Rutherford in Corry 1979)

When the torpedo was released, the heading and position of the plane and ship were recorded to enable the instructor to calculate whether a hit had been scored. On some exercises the ship and/or the pilot would take evasive action after the torpedo was dropped.

At this stage in the development, the TAT specification was for a simulation of the final straight and level approach, followed by the launch of the torpedo at the target. As the prototype was being constructed, a demonstration flight of a torpedo attack using a swordfish revealed that this was inadequate as the dive from height followed by turn(s) to line up on the target was an essential part of the attack and that these must be included in the TAT. With the risk that the project would be cancelled without these vital elements of the simulation, the necessary modifications were developed (Corry 1979).

A dive computer was added; this enabled a modified epidiascope to vary the view of the ship depending on the height and distance of the aircraft (Humphris & Sons Ltd. 1944). Unfortunately it was impossible to build these into the prototype, due to the extensive reworking required, but was included in subsequent installations. Other changes included the cyclorama being extended to 360°, with two windows for the instructors, and an access tunnel for the pilot. The floodlighting was simplified and some sound effects were introduced.

How the Virtual Environment Was Achieved

Banks of floodlights arranged in an arc provided a coloured background. Initially triplets of floods were used, coloured red, green and blue (Rose 2002), so that any colour (in theory) could be chosen. Later specifications (Air Ministry 1946) have only two sets, each a different shade of blue suitable for sky effects. This change enabled a better sense of realism, since additive mixing of primary colours using the filters of the period often only gave poor results for non-saturated tones.

The various clouds, depending on the simulated weather required, were projected using large format slide projectors, one for each cloud. The slides were hand-painted in black and white and the resulting images coloured, as appropriate, using gelatine sheets in normal theatre practice. The projectors were mounted in an arc, matching the floodlights, with the trainer at the centre.

A seascape, with horizon, was painted on the cyclorama, when the aircraft was below 500′; this was enhanced by projecting moving waves, whose speed was controlled by the instructor, onto the cyclorama below the horizon. Appropriate ambient lighting of the aircraft was provided as well as a sunset effect (Air Ministry 1943).

All of the lighting was push button-controlled by an automatic bank of dimmers, from which the instructor could select, for example, bright night or overcast day, and all the lights would fade automatically to the appropriate level. This was achieved electromechanically with cams, gears, levers, relays and switches.

In order to project the image of a ship, turning and moving in respect to the virtual motion of the Link Trainer, where the view of the ship's deck would realistically change with the simulated height of the aircraft, was far beyond anything that had been attempted before. The projector was required to both pan and tilt and for the image to zoom to enable the ship to move and grow as the plane approached. To achieve this, an epidiascope (a projector that projects an image from a real 3D object) was developed with the assistance of Strand Electric at their demonstration theatre. The initial optical design was completed by much trial and error. Unfortunately the demonstration theatre was bombed the evening it was completed, 11 May 1941 (Strand Electric 1941); the prototype survived, but it took 4 weeks before that part of the building was safe enough for it to be recovered (Rose 2002). The epidiascope projected a 1:1200 (1″ to 100′) scale brass model of the enemy shipping. These were lit with four banks of lights controlled from the push button controller like the rest of the lighting. Thus the target could be evenly lit, like daylight, as if at night, under a searchlight, or with the addition of an automated flip-up screen, silhouetted against the horizon.

Although common today, none of these features were automated on period projectors and did not become common in the entertainment industry until the early 1980s.

Development of Other Simulators

With the success of the TAT at Crail, more were ordered by the Fleet Air Arm and the Royal Air Force. Other branches of the Navy were keen to use similar techniques for the training of torpedo attacks. The main development was that for a combined destroyer/Motor Torpedo Boat (MTB) trainer. This had very similar projection technology to the aircraft version, and the instructor's units were supplied by Link as before. However instead of the aircraft, a mock-up of the destroyer's bridge structure was constructed. This was still equipped to turn, pitch and roll, and the motion was realistic enough to induce seasickness in some of the trainees (Fig. 9.2). To maximise the usefulness of these trainers, the MTB controls were mounted back to back with that of the destroyer, so a change from one vessel to the other merely involved the bridges to be rotated through 180°.

TAT: Submarine

The submariners insisted that a similar trainer be developed for their service. After the initial specification problems with the air version, submarine torpedo training off the Isle of Arran was observed prior to commencement of the project. As a result a reproduction control room complete with access via conning tower was created,

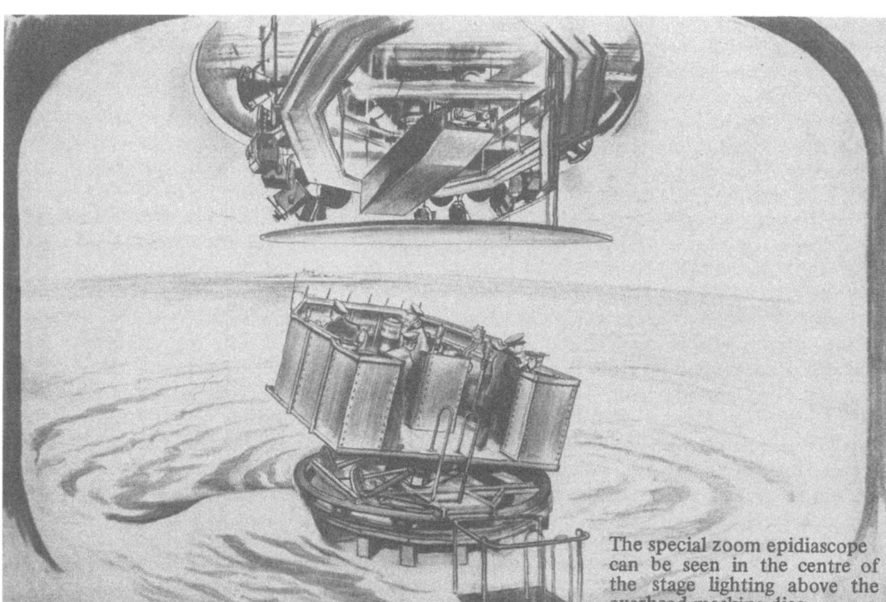

Fig. 9.2 Destroyer/MTB. Combination. TAT – Surface Craft. (Rutherford, in Corry 1979)

and a special periscope to observe the targets was developed by the Glasgow firm, Barr & Stroud. The facilities to provide the virtual battle environment were similar to that for the destroyer/MTB TAT.

Operational Crew Trainer

The 'terrain' of most torpedo attacks is comparatively straightforward to simulate: open sea, weather and variable visibility. The final simulator developed the Operational Crew Trainer (OCT) aimed to simulate a large seascape, as viewed from several thousand feet away. Relief maps showing the topology of the landscape have been used in war since at least the sixteenth century (Warmoes 1999), whilst photo mosaics were used extensively in the First World War (Finnegan 2011). The OCT went far beyond either of these in terms of its interactivity and thus the relationship of the aircrew with the landscape.

At Machrihanish, a former site of an aircraft TAT, a radically different trainer was built. It was designed for training pilots and observers for directing shore bombardments from 'large warships'. This, however, was questioned by some of the pilots whether this activity was worthwhile with the number of battleships already destroyed by the time it was operational (Ough 1999). The OCT was different in many regards to the earlier TAT. It could accommodate up to four aircraft crews simultaneously each in their own 'aircraft', and each was mounted in dedicated cubicle with a window onto the main target area, thus isolating them from each other. The target area consisted of an approximately 180° screen upon which the usual range of sky effects could be projected. A non-rotating epidiascope enabled an additional target to be projected at the centre of the screen. From Rutherford's (in Corry 1979) sketch (Fig. 9.3), it appears that an additional projector could be also added at ground level on a trolley.

The biggest difference between the OCT and the TATs was the main target area. This consisted of a 30′ diameter revolving stage on which a scale model coastline and harbour installation were built so it appeared as if it was several miles distant, from a height of approximately 3000′. In order for the naval bombardment to be realistic, a number of features were included in the model. A large number of small lamps, which would simulate the flash of an exploding shell as it landed, were secreted in the artificial terrain. Small electrically driven pegs allowed sections of the landscape to suddenly rise up, as if hit. Some of the buildings had hinged sections, which could be driven by pegs allowing walls to fall over due to blast. A mechanism was even included for splashes of water for those shots that fell short into the river or sea. The final special effect was the dropping of parachute flares (at night) over the target which was achieved by slowly lowering small lamps over the model landscape (Corry 1984). The Link Trainers, used in the OCT, unlike in the other TAT, were not fully automated, although they could be rotated with respect to the window, and thus the target. The speed of rotation of the revolve could be tied to the simulator's airspeed so the landscape moved past realistically.

Fig. 9.3 Operational Crew Trainer at Machrihanish (Rutherford, in Corry (1979))

The other large-scale synthetic trainers of the period to introduce the battle space interactively were the Link Celestial Trainer (British Air Commission 1941), which included a realistic (by time and position) night sky for navigation training, and the AML Teacher (Fellowes 1942). Both relied on large photo mosaics for their target areas which moved with the motion of the aircraft but were otherwise noninteractive.

Comparison of TAT (Aircraft) with Modern Simulators

Modern flight simulators are classified by the Federal Aviation Authority (FAA 2008) and other national aviation authorities for example Transport Canada (2010), using similar criteria, in order to differentiate the level of realism and thus their utility in flight training. These vary from unclassified, computer game 'simulators', through category A which are basic motion simulators to category D which fully simulate the flying environment (Table 9.1). Their use reduces the hours trainees require in the air and also enables the rehearsal of emergency procedures which are not possible to safely recreate during actual flight.

Communication and controls used in the TAT would be adequate for a modern simulator. Ground handling was entirely absent in the TAT, whilst the responsiveness

Table 9.1 Summary of FAA Simulator Classification (after FAA 2008). Each level includes the characteristics of the previous categories

Category	General features	Visual systems	Motion systems
A	Replica cockpit, including operational communications, instruments, circuit breakers Aerodynamic drag modelled Sounds resulting from pilots actions simulated Response time 300 ms Ground operations generically represented	A 45° horizontal and 30° vertical field of view for each pilot Visual scene content at a decision height on landing approach	At least four degrees of freedom Force cues shall be perceived by the pilot
B	Ground handling and increased aerodynamic programming	The visual system shall provide cues to assess sink rate and depth perception during landing Test procedures on visual system	Special effects for specified buffets and bumps
C	Control feel dynamics to match actual aeroplane Brakes (and skids) modelled Wind shear modelled Other aeroplane sounds simulated (including crash for over heavy landings) Response time 150 ms	A 75° horizontal and 30° vertical field of view per pilot or 150° cross cockpit horizontal view Night and dusk simulation Minimum resolution and brightness standards	At least six degrees of freedom
D	Full aerodynamic modelling All sounds at appropriate amplitude	Daylight, dusk and night visual scene content for recognition of airport, terrain and major landmarks	Characteristic buffet motions of the simulated aircraft

of the systems is impossible to evaluate without a working example. However ongoing research is attempting to recreate elements of the TAT from drawings and original components to further understand the systems. The visual system gives a 300° horizontal and 46° vertical field of view, night and dusk are simulated, and most of the resolution requirements are inappropriate due to the analogue nature of the projections but would not suffer the pixelation issues of some modern digital systems. The TATs were not designed to provide landing training, so obviously they don't meet those requirements; however the needs of the dive and final low-level approach have similar characteristics. The motion system in the TAT has four degrees of freedom, and at least some of the force cues are felt by the pilot, and, although not specific to any particular aircraft, buffeting is available.

Within the limits of the design scenario – to simulate the dive, approach and torpedo launch – the TAT had a performance that measured up well against basic simulators of today. The TAT provided an excellent field of view with a range of day and night scenes with varying weather options. However motion is limited to four degrees of freedom; the aerodynamic modelling was basic, with only limited stall buffeting; and sounds were minimal.

Conclusion

The TAT provided a level of realism in their simulations which surpassed anything previously possible, and although they did not meet the standards of many of today's simulators, they were astounding in their capabilities. The combination of technologies to perform the necessary calculations to control the aircraft, target ship, projectors and interactive landscape of the OCT was truly creative. Today of course a digital computer would be used, but the necessity for the FAA to specify reaction times shows this is not necessarily trivial even today.

The TATs success could also be measured by the 60 simulators built (Strand Lighting 1945) which was only exceeded by one other simulator including a the virtual battle space, of the period – the Link Celestial Trainer, which had hundreds constructed – but the RAF returned most of their 60 celestial trainers under reverse lend-lease. The base Link Trainer utilised by both was made in the thousands.

Their success is perhaps best measured by the view of the pilots; John Ough, who 'flew' the OCT, said:

> The realism was amazing. After a few minutes of turning this way and that in the dim, muted light, the illusion of actually flying became very real. (Ough (1999))

Even with the changing military roles and focus postwar that some of the trainers remained in use showed their utility and continued role in training military personnel. Examples of the destroyer/MTB models were exported to Sweden (Corry 1984) and apparently Norway (Admiralty 1952), demonstrating a wider appreciation of this type of training.

References

Admiralty. (1952). *Estimates and conditions for supply and erection of destroyer MTB type torpedo attack teacher (CYCLORAMA) for the Royal Norwegian Navy delivery arrangements etc. [document] ADM 1/21532*. London: National Archives.

Air Ministry. (1943). *Specification of wiring installation and description of lighting apparatus for T.A.T./A2*. London: Air Ministry.

Air Ministry. (1946). *Air Publication 2655C*. London: Air Ministry.

Bentham, F. (1992). *Sixty years of light work*. London: Strand Lighting Ltd.

British Air Commission. (1941). *[document] 21.7.1941 AVIA 38/769*. London: National Archive.

Corry, P. (1979). *Theatre at war. Sightline 13(2):1108*. London: Association of British Theatre Technicians.

Corry, P. (1984). *Our world was the stage*. Manchester: Neil Richardson.

FAA. (2008). 14 CFR Part 60 (2016) [online]. Available at: https://www.faa.gov/about/initiatives/nsp/media/14CFR60_Searchable_Version.pdf. Accessed 27 Nov 2016.

Fellowes, P. F. M. (Ed.). (1942). *Britain's wonderful air force*. London: Odhams Press Ltd.

Finnegan, T. J. (2011). *Shooting the front* (p. 164). Stroud: The History Press.

Flight. (1910). *The Stanley Air Show Flight 2(47):951*. London: Flight Publishing Co. Ltd.

Flight. (1939). *J.V.W. Corporation Ltd. Flight 36(1615):474*. London: Flight Publishing Co. Ltd.

Haward, D. M. (1910). *The Sanders "teacher" flight 2(50):1006*. London: Flight Publishing Co. Ltd.

Humphris & Sons Ltd. (1944). *Dive Computer General Assembly dwg. [document]*. Parkstone: Humpris and Sons Ltd.
L3 Simulation and Training. (2012). [online] Available at: https://www.link.com/about/pages/history.aspx. Accessed 19 Mar 2013.
Ough, J. (1999). *Crumbs!* Burnstown, Ontario: General Store Publishing House.
Rose, P. (2002). Tales from SEECOL. Focus June / July 2002, Association of Lighting Designers, Southsea.
Strand Lighting. (1945, December). Theatrical lighting. London: Strand Electric and Engineering Co. Ltd.
Strand Electric. (1941). *Minutes of the Board of Directors Bentham Collection THM/46/5/6*. London: Victoria and Albert Museum.
Transport Canada. (2010). [online] Available at: http://www.tc.gc.ca/eng/civilaviation/publications/tp9685-chapter2-appendix2a-1047.htm. Accessed 21 Mar 2013.
Warmoes, I. (1999). Musée des plans-reliefs: Historic models of fortified towns Caisse Nationale des Monuments Historiques et des Sites Éditions du Patrimoine Paris.
Williams, B. W. (1929). Pre-war stuff. *Popular Aviation, 4*(1), 66.

Chapter 10
The 1968–1973 Egyptian Army Field Preparations for Crossing the Suez Canal and the Conflict Between Israel Defense Forces Intelligence Research Units

Joel Roskin and Eliyahu Dekel-Dolitzky

Introduction

Geographic Intelligence and Balancing Between Intelligence Disciplines

There are many connections between geography, terrain, space, and military action and state security. One of them is the spatial pattern of defensive and offensive security and military infrastructures (Oren and Roskin 2015). Geospatial intelligence (GEOINT) is an emerging and computerized-based discipline that has evolved from VISINT – visual intelligence. GEOINT deals with the connections between geographic features and security issues by providing detailed spatial data analysis to assess the operational environments of opponents. This is done by exploiting and analyzing imagery and other geospatial data sources to describe, assess, and visually depict natural and human-made physical features and geographically referenced activities (Sanchez 2009).

Geographic intelligence is often accomplished by people who have a technical, quantitative, and scientific orientation, such as aerial photograph interpreters, spatial analysts, and geographers, and requires a less humanistic approach than

J. Roskin (✉)
Analytical Laboratory, Excavations, Survey and Research Department, Israel Antiquities Authority, Jerusalem, Israel

Geomorphology and Portable Luminescence Laboratory, Leon Recanati Institute for Maritime Studies, University of Haifa, Mt. Carmel, Israel
e-mail: yoelr@bgu.ac.il

E. Dekel-Dolitzky
Ely Dekel Books, Elkana, Israel
e-mail: elyahu-d@bezeqint.net

P. L. Guth (ed.), *Military Geoscience*, Advances in Military Geosciences,
https://doi.org/10.1007/978-3-030-32173-4_10

other intelligence research. Geographic intelligence is often merely intended to supply basic field data from a perspective of modeling and visualization; it is not considered appropriate for detailed spatial information collection and research (Sanchez 2009). It is thus often perceived as a minor component of intelligence assessment. This incomplete understanding of the capabilities of GEOINT and the tendency of intelligence agencies to strongly focus on traditional signal source intelligence (SIGINT) and human-sourced data (HUMINT) often retards exploitation of GEOINT in operational planning.

A balance of information collection and analysis between different intelligence disciplines is required to construct a holistic intelligence assessment, but this approach is usually controversial. For example, recent rapid advances in technology have downplayed the relative importance of classical espionage (Barrowman 2007) and have led to a surplus of confidential and open information collection in all the intelligence agencies. Analysis and interpretation of the tremendous volume of terrain data often requires a technical focus that downplays research of such data (Clark 2013).

Tension between geographic intelligence and "research intelligence," i.e., SIGINT and HUMINT, is common in many agencies and is a potent source of producing incompetent intelligence that often limits the success of consequent operational results (Roskin 2018). These interorganizational rivalries have not been highlighted by research intelligence units and in discussions, and examples of this organizational situation and its outcomes have remained behind closed doors. In some agencies, geographic intelligence is not even regarded as intelligence.

Interorganizational tension can be attributed to several causes: (1) the lack of personal and professional military and intelligence field experience of SIGINT and HUMINT personnel; (2) the mindset that technical applications of imagery products should be used just for visualization, maps, and target aids and details; (3) human-based intelligence due to the understanding that humans are the initial initiators and developers of war and that terrain is only a relatively stable platform upon which war takes place; and (4) limited access of geographic intelligence personnel to other types of intelligence data, mainly due to confidentiality of sources and interorganizational prioritizing (Dekel-Dolitzky 2010).

The surprise crossing of the Suez Canal (SC) by the Egyptian Army initiated the outbreak of war between Egypt and Syria against Israel on October 6, 1973. The offensive occurred on the holiest Jewish holiday and fast day, Yom Kippur (the Day of Atonement), and it is thus known in Israel as the Yom Kippur War. In Egypt, the War is known as the October War and is perceived as a victory because Egyptian forces successfully crossed the SC. In this paper we will use the term 1973 War. The surprise, perceived as an intelligence fiasco by the Israel Defense Force (IDF), has received much attention (Bar-Joseph 2012 and references within), but the process and challenges of geographic and visual intelligence analysis, based upon aerial photograph interpretation in predicting the Arab offensive (Shlaim 1976), have not been presented.

Aim of this Paper

This paper is based upon the 2nd author's personal account who served as an officer in the S1 unit and classified intelligence publications that he published or read at the time of the described events. The paper chronologically reviews the 5-year process of the IDF geographic intelligence research through the stages of the Egyptian preparation of infrastructure for crossing the SC. These began around 1968 and ended with the onset of the 1973 War. The identification and analysis of step-by-step tactical terrain changes involved in preparations for the SC crossing are presented. This study also reveals a rare look inside the personal and organizational processes of collection and analysis of geographic intelligence by the IDF's main geographic intelligence unit known as "Section 1" (S1). Herein the term "geographic intelligence" is defined as intelligence based on research of military geographic features collected from GEOINT sources.

Background

Geography and History of the Suez Canal

The Suez Canal, constructed in 1859 by a French company, crosses the northeastern Nile Delta and the northwestern Sinai Peninsula. The Sinai-Negev sand sea stretches to the east (Fig. 10.1; Roskin et al., 2014; Roskin and Tsoar 2017). As such, its central and northern parts are within sand deposits, whereas its southern part is within finer sediments that are easier to consolidate: clay and silt (Hoffmeier and Moshier 2006). The SC links the eastern Mediterranean Sea to the Red Sea via the Gulf of Suez and provides a strategic connection between the shores of Europe, Africa, and Asia (Karabell 2003).

The SC is in total 162 km long and connects three natural, shallow lakes; the excavated sections total 112 km. Originally, the width was 45 m, but over time, it was widened, expanding to the west, so that in 1973, it was 160–180 m wide. The narrower Sweet Water Canal (SWC) runs parallel to and west of the SC and comprises a minor water obstacle. Completed in 1863, this canal was designed to supply local drinking water (Hoskins 1944).

The SC is both a strategic financial asset and a substantial military obstacle for maneuvering forces into the Sinai Peninsula and as such has generated substantial political and international attention. The Convention of Constantinople in 1888 declared the SC a neutral zone under the protection of the British. The British were later to defend this strategically important passage from a major Ottoman attack in 1915 during World War I. Under the Anglo-Egyptian Treaty of 1936, the United Kingdom (UK) insisted on retaining control over the SC. In 1951 Egypt repudiated this treaty, and in 1954, the United Kingdom agreed to remove its troops from the Canal (Karabell 2003). The Egyptians aggressively controlled the Canal to include

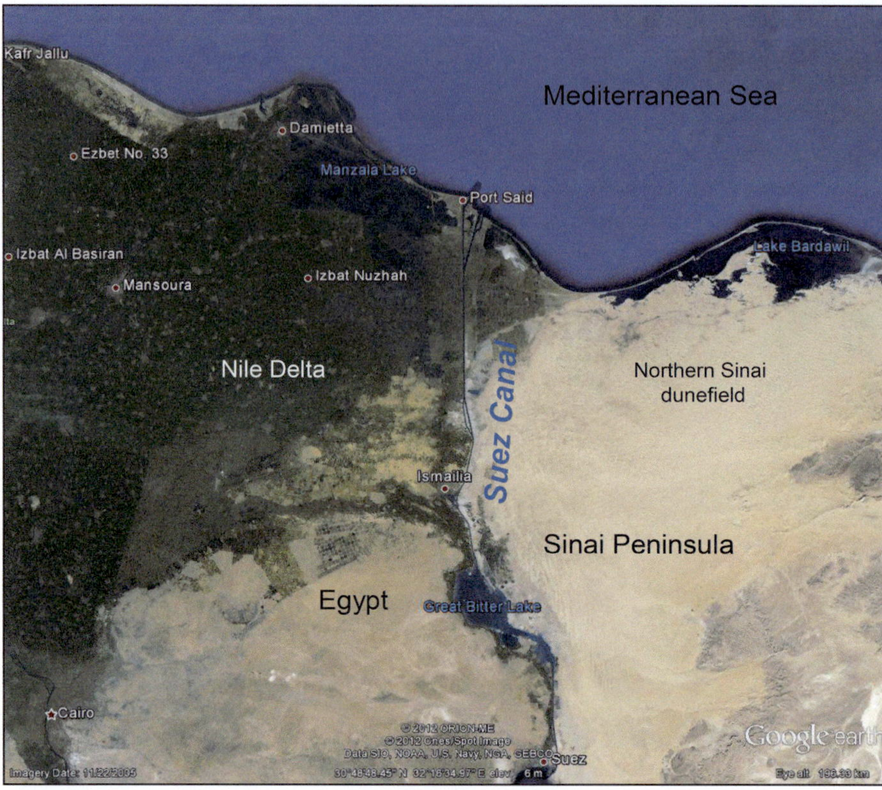

Fig. 10.1 (**a**) Google Earth image of the eastern Nile Delta, the Suez Canal, and the northwestern Sinai Peninsula

blocking Israeli shipping. During the Six-Day War between Israel and its Arab neighbors in 1967, the IDF destroyed large parts of the Egyptian Army and Air Force, conquering the Sinai Peninsula, and seized control of the eastern bank of the Suez Canal (Oren 2017).

Egyptian Fortification Along the Suez Canal

The Egyptians dealt with crossing and maintaining the SC in various fashions. Before the Six-Day War when the SC was under Egyptian control, vehicles crossed the Canal on civilian ferries and military pontoons. When a military force had to cross the waterway, commercial sailing was halted, and bridges were assembled and installed.

To limit bank erosion, the Egyptians laid a rock foundation, essentially L-shaped cement cubes, at a depth of 2 m, and above the waterline, they built a 1.5-m high and

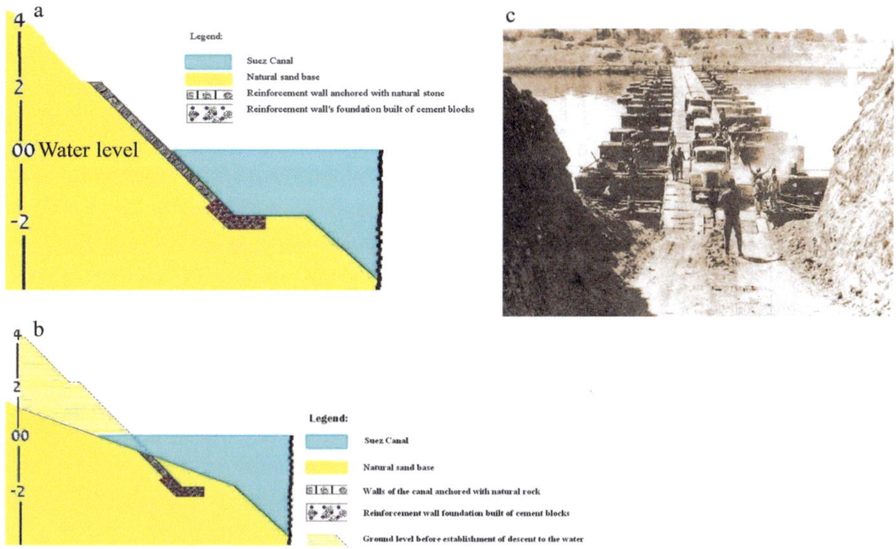

Fig. 10.2 The western bank of the Suez Canal (figures after Dekel-Dolitzky 2010). (**a**) Reinforced slope of sand embankment with natural rock (I) and cement block reinforcements (II), (**b**) Modification of the west bank for military (descents to the water) access to the water line, (**c**) Egyptian 1973 breakthrough in the Bar-Lev Line embankment and floating bridge

vertical stone wall, and, finally, above that, a retaining wall that sloped at a 30–45^0 angle was constructed. However, the cement cubes and incline of the retaining wall proved to be an obstacle for massive vehicle access to the SC western bank required for an offensive canal crossing (Fig. 10.2a, b). These obstacles influenced the nature of the Egyptian preparations for the SC crossing in 1973 as described ahead.

War of Attrition and the Egyptian Response to the "Bar-Lev Line"

During 1969 and 1970, Egypt carried out a "War of Attrition" against IDF troops deployed on the east bank of the SC that mainly involved artillery and sniping fire across the SC. These small-scale offensives were part of the reason the IDF intelligence and command agreed that only such minor engagements were within Egyptian military capabilities and not full-scale war. Nevertheless, five Egyptian infantry divisions of the 2nd and 3rd field armies were deployed on the western bank of the SC including 2200 tanks and 2000 artillery pieces during this period (Dekel-Dolitzky 2007; Dekel-Dolitzky 2010).

In response to the War of Attrition shelling, and as part of a long-term defense plan, the IDF began to construct at the end of 1968 a defensive layout that included, inter alia, underground strongholds for soldiers connected by an embankment obstacle along the eastern SC bank – the "Bar-Lev Line." The Bar-Lev embankment and strongholds went through several stages of development that resulted in a continuous embankment 15 m high and ~50 m wide. The embankment was made out of local sediment: sand in the north and silt in the south. The Line was thought to be impassible for an Egyptian SC crossing by the IDF command, unless abundant and easily seen heavy engineering equipment was to be deployed (Bar-Siman-Tov 1988; Dekel-Dolitzky 2010) (Fig. 10.2c).

The Bar-Lev Line led to an asymmetric fortification situation where the Egyptians relied on their short vertical rock and retaining wall embankments which were strikingly inferior to the Bar-Lev fortifications. This situation enhanced the confidence in the IDF that the Egyptians will not attack nor plan an offensive crossing of the SC. Furthermore, it forced the Egyptians to update their initial post-1967 plans and capabilities for crossing the SC and explore practices to overcome the trafficability and observation constraints that the Bar-Lev Line inflicted. Furthermore, the linear array of Egyptian military construction along the SC was minute, consisting of reinforcement made of modular iron hoops. Some fortifications also were built on the embankment of the SWC. These small-scale structures reinforced the perception of military superiority within the IDF command (Fig. 10.3).

Fig. 10.3 The Sweet Water Canal. (**a**) Sketch of a submerged underwater passage made of gravel, (**b**) Egyptian armored personnel carrier stuck on a submerged passage. (Photo by E.D)

IDF Intelligence Section 1 Organizational Structure and Geographic Intelligence Research Methods

The Egypt-oriented branch of "Section 1" (S1) that specialized in aerial photograph interpretation and geographic intelligence was initially divided into three teams commanded by majors (Fig. 10.4). These were an in-depth infrastructure frontline (SC region) team, operations teams, and targets teams. The depth infrastructure frontline team, headed by the late Benny Kain, dealt with the Egyptian field armies deployed on the western bank of the SC. Kain was an economist and businessman, who enlisted as a career officer following the Six-Day War, and later served as a reservist and a part-time civilian worker. The targets and operations teams dealt with immediate operational needs, and the long-term and less urgent infrastructure research of the SC region was handled by Kain and his team.

S1's main collection source was aerial photography from a Hycon 73B, Lockheed U-2C aerial camera designed for U2 high-elevation reconnaissance planes. The 36 in. cameras were designed to cover the underlying terrain from horizon to horizon and resolved features as small as 0.5 m. The cameras were adapted and mounted on F-4 Phantom combat planes. The film was developed on a positive celluloid roll and, by means of a simple light table and later by a MIM4 (light table) deciphering device, was scanned and interpreted in a magnified stereoscopic mode (Colwell 1985; Brugioni 1996). Generally, the film was sent upon arrival to an aerial photo interpretation team to perform a quick scan to identify military vehicles and surprise attack threats. Only later were the film and copies distributed to the depth infrastructure frontline team who conducted geographic intelligence research.

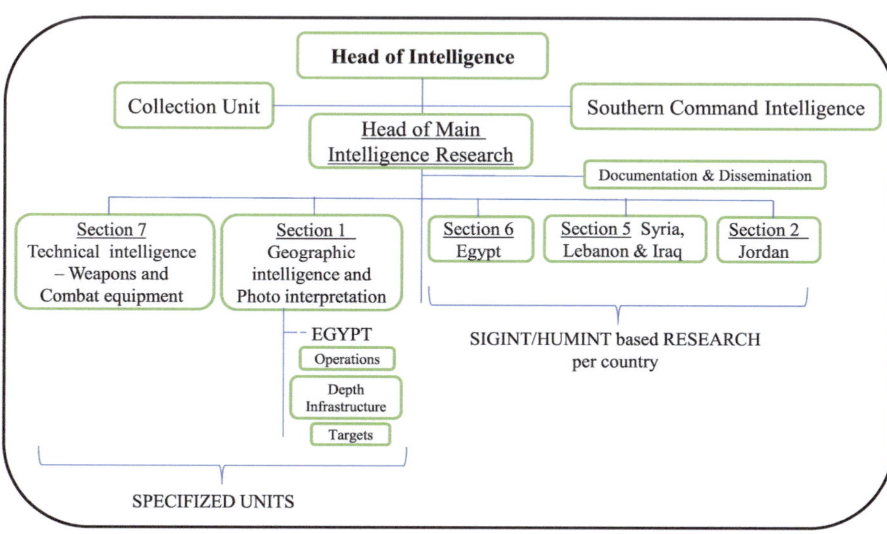

Fig. 10.4 Organizational chart for IDF intelligence units discussed in this paper

Until the end of the War of Attrition, many cross-border sorties were made above the west bank of the SC, and vertical and high-resolution photos over the area were obtained. In August 1970, following Soviet intervention and the positioning of anti-aircraft missiles along the west bank, a cease-fire was agreed. After this, Israeli Air Force (IAF) reconnaissance photography was restricted to flying over the west bank of the SC, and thus oblique photos became the main remote GEOINT source. Oblique photos made interpretation a tedious task because the distance covered from top to bottom in any given image was greater than for a vertical photograph and the angle of view decreased with distance (Colwell 1985). Analysis was sometimes based on comparison with previously interpreted spatial patterns and not on feature details in the image, which could be unclear, and dimensions.

Geographic Intelligence Concerning Egypt's Plans for a Canal Crossing

Background

In 1968, based on the quantity of the structures and the quality of the amassed military vehicles that his team identified, Kain determined that the Egyptians were planning a massive Suez Canal crossing. He was also the first to begin tracking their progress methodically and rigorously. However, the time for materialization was unclear, and SIGINT- and HUMINT-sourced evidence for this hypothesis was lacking.

The lack of supportive intelligence motivated the depth infrastructure frontline team to identify additional physical signs of imminent war to convince other intelligence sections and the higher echelons that their interpretation was correct. This effort produced five main geographic intelligence discoveries before, during, and shortly after the War of Attrition: road networks, underwater passages on the SWC, road descents and crossing platforms, parallel-pair assault revetment construction, and assault revetments for canal crossing equipment. These were supported by an additional five types of finds between the War of Attrition and the 1973 War. The latter, described below, led to a comprehensive assessment within S1 that an Egyptian operational plan to cross the SC was rapidly maturing.

Road Networks

In 1968, Kain had already discovered a new network of dirt roads that converged into a single road close to the Sweet Water Canal (Fig. 10.5). This road led to the banks of the SWC where a new bridge was under construction, from where another dirt road was being laid right up to the west bank of the SC. Kain concluded that the roadworks and the new bridge meant that a battalion or even brigade-sized crossing

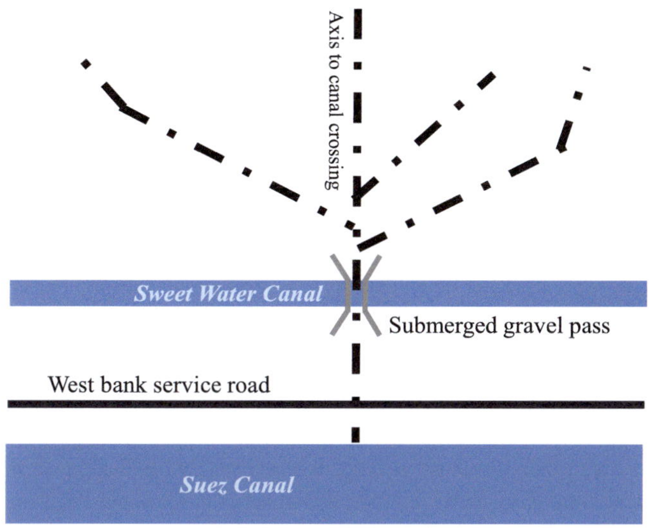

Fig. 10.5 Sketch of a dirt road network that converged into a single road close to the Sweet Water Canal

was being prepared. He reasoned that if the infrastructure was intended to serve a smaller force, such as for a raid or intelligence gathering mission, the Egyptians would not need so much infrastructure.

Kain's discovery, confirmed by the S1 head and disseminated in an intelligence report, caused a stir in the intelligence management. The Egyptian Army had been defeated in 1967, and it was expected by the IDF intelligence management that it would require many years to recover from this defeat. It was thus inconceivable that the Egyptians could manage to embark upon preparations for an offensive crossing the SC in only a year. The IDF Southern Command's intelligence unit that had operational responsibility on the SC and the military subsection in the Egyptian section of the main intelligence research department (S6) rejected Kain's interpretation out of hand. They were so confident that S1's interpretation was wrong that they did not task the collection unit to verify it (Fig. 10.4). Lacking SIGINT- and HUMINT-source confirmation, intelligence staff believed that S1's conclusions were unreasonable and unprofessional and that Kain himself was unreliable. However, the head of S1 fully supported Kain's assessment and instructed him to search for more indications that the Egyptians were indeed planning a canal crossing with a large force.

Underwater Passages on the Sweet Water Canal

Shortly after the discovery of the initial road network and bridge at the SWC, more bridges were found in other sectors to the north and south along the Canal. This further indicated to the researchers in S1 that a comprehensive strategic plan was unfolding. While the bridges were being constructed, Kain discovered other places where roads extended in the direction of the SWC were being paved but with no bridge being built. At first it was thought that the bridge parts simply had not yet arrived, but when no further work occurred, Kain determined that a new type of bridge, what he termed an "underwater passage," had been developed. Pipes were laid on the canal bottom to allow a continuous flow of water, then the infrastructure was covered with sediment and stones, and the upper section of the "bridge" was immersed. A vehicle could then easily cross the water obstacle (Fig. 10.3a, b). This observation was also interpreted from aerial photos by the terrain branch of the Southern Command intelligence but was not accepted by the Command's directorate. Kain then determined, from a time series of aerial photos, that construction material that had not been there earlier had been stockpiled at the bridge sites and then had "disappeared." This suggested to him that the material had been deposited in the SWC to serve as the base for an underwater vehicle passage.

Again, this discovery elicited antagonism in the intelligence command as supportive SIGINT and HUMINT data were absent. With time, however, Southern Command directorate accepted the underwater passages discovery, and the main intelligence establishment began to understand that the Egyptian Army was indeed creating infrastructure designed for a Suez Canal crossing.

Road Descents and Crossing Platforms

After the Egyptians overcame the SWC obstacle by bridging and underwater passages, they turned to the western bank of the SC. As mentioned above, the cement and stone cubes on the Canal bank, and especially its reinforcement wall, constrained access to the SC for vehicles and amphibious equipment. The Egyptians used bulldozers to break through the low embankment and reinforcement wall to create accessible "descents to the water" for amphibious equipment (Fig. 10.2b). These breakthroughs were prepared openly in front of IDF troops on the opposite bank of the Canal. After the descents were completed, the Egyptians levelled platforms between them on the Canal bank, each of which was dozens of meters in length; these were termed "crossing platforms." Built between the descents to the water, it was thought that the platforms were intended to serve as assembly areas for bridge sections that would be thrown into the Canal once assembled. It was also assessed by the depth infrastructure frontline team that the descents were to be used for infantry assault boats and for dropping the advanced, rapid-assembly, Russian-made PMP floating bridge equipment into the Canal (Fig. 10.2c).

At first the physical evidence and the implications of the construction work that a SC crossing was in the making was unchallenged. The "brazen" Egyptian construction operations thus confirmed Kain's earlier assessments. However, with time, denial and counter-speculation returned and dominated among non-geographic intelligence staff. The numerous geographical observations still did not fit the general joint assessment made by intelligence and high echelon generals because there were no HUMINT reports even hinting at such intentions. In addition, visual intelligence of an array of tactical finds was far from reflecting the operational magnitude and the complexity of crossing the SC. Intelligence command continued to insist that the Egyptians lacked such operational capability so soon after their 1967 defeat. Furthermore, it was also assumed that an Egyptian attempt to cross the SC would be easily thwarted by the IDF.

Parallel-Pair Assault Revetment Construction

While the descents to water were prepared along the entire length of the Canal, Kain identified a new type of ground work being carried out 5–10 km west of the SC. Pairs of parallel earthen berms were being constructed by heavy mechanical equipment (Fig. 10.6). The parallel berms, separated by 4–5 m, were 30 m long. Groups of 30 berm pairs at 30–40 m intervals were identified along the SC in the 2nd and 3rd (Egyptian) field army sectors. Ground work of this type had never been seen before. Kain believed the berm pairs were meant to serve as protected areas for vehicles transporting the bridging and crossing equipment. However, the grand plan behind this engineering setup was unknown.

Protecting these vehicles by revetments was necessary for Egyptian units because during the War of Attrition, the IAF had executed thousands of offensive sorties. The intelligence challenge for this discovery was that the previous vehicle shelters were three-sided, U-shaped earthen embankments, only 5–10 m long. The U-shaped revetment geometry was assessed to possess a drawback: a vehicle entering or exiting the dugout had to reverse and could thus possibly jam traffic or cause an accident. Kain hypothesized that the new parallel-pair revetments were designed for rapid entry and egress and that their 30 meter length provided protection from shelling from their sides. He termed these new constructions "entry and exit revetments" and "assault revetments." He explained that the huge amount of bridging equipment, upon which any SC crossing operation depended, involved hundreds of trucks loaded with bridging sections and various other types of fording

Fig. 10.6 Parallel-pair assault revetments

equipment. The vehicles would be exposed to attack and had to be protected for hours until the equipment was dropped into the water and bridging operations could begin.

Assault Revetments for Canal Crossing Equipment

Beginning in 1970, Kain's team had discovered new concentrations of assault revetments, 120–150 of them in each concentration, tens of kilometers west of the SC, despite blurry imagery. Kain understood that they were meant to serve as "entry platforms" for crossing units before they moved into the combat and crossing zone beside the SC. The interpreted operational concept was that engineering units would pull the bridging equipment out of pre-existing storehouses, such as in Wadi Hof, south of Cairo, and deploy it on the entry platforms to the rear of the field armies until the order came to commence crossing operations.

Despite requests by Kain's team to acquire more information on the assault revetments, the IDF intelligence collection units did not supply any supportive intelligence for these finds and assessments to S1. Collection explained that due to the confidentiality of their sources, they could not pass on such intelligence to terrain analysis units. Indeed, huge stockpiles of crossing equipment were later found by the IDF ground troops in Wadi Hof during the 1973 War.

Pre-1973 Discoveries of Egyptian Engineering War Preparations

Tank Ramps on the Canal's Embankments

In mid-1972, the Egyptians began raising embankments at many places along the western bank of the SC to a height of 20–25 m, higher than the Bar-Lev Line embankments (Fig. 10.7). Kain claimed that these higher embankments whose western ascents were at an angle of about 20° were intended as sites from which tanks (tank ramps) would provide covering fire for the SC crossing units at the start of a war. Since the slope of the eastern face of the embankment (as viewed from the east bank) precluded the descent of tanks, Southern Command intelligence interpreted them as positions for Egyptian spotters and tank hunter teams equipped with AT-3 Sagger anti-tank missiles that could be useful if the War of Attrition reignited.

During the 1973 War, the positions were utilized for both tanks and Sagger squads. The adjacent tank cover and the protection of the SC obstacle from Israeli ground attacks made the squads operate more comfortably in relation to Sagger squads that ambushed IDF tanks along roads within Sinai.

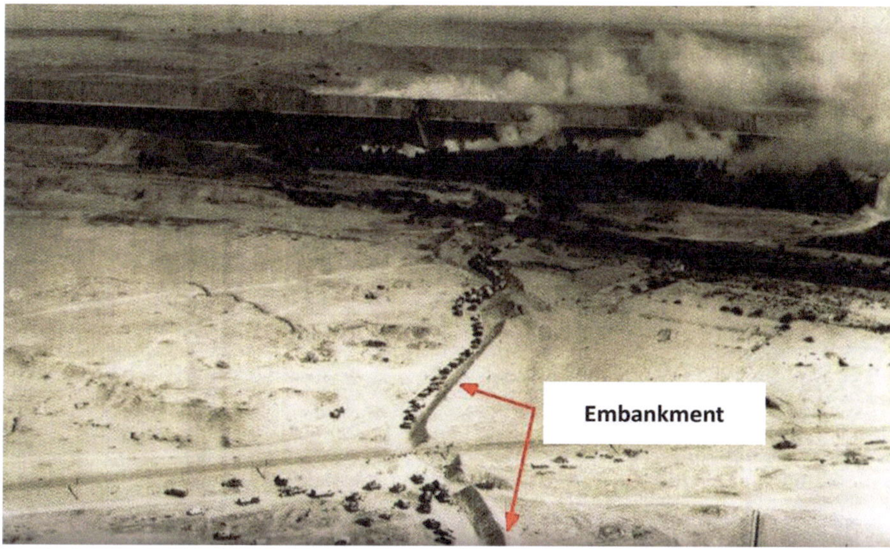

Fig. 10.7 Oblique image showing earthen embankments along the access roads leading to the Suez Canal and the Bar-Lev embankment in the background

Drinking Water Pipe Layouts

When engaged in routine defense and training, the Egyptian Army water supply was based on tankers. However, for crossing the SC, numerous tankers that could jam the traffic and hinder the crossing would be required. To solve this problem, a water pipe was placed along western bank service road of the SC in 1972, and cement water reservoirs and pumping stations were established.

S1 identified by aerial photographs the trenches and reservoirs. The S1 staff with the assistance of a water expert studied the matter and circulated an intelligence review that stated that the new water system could serve the Egyptian Army in case of a renewal of fighting or in the event of a massive SC crossing.

During the 1973 War, the Egyptians applied special engineering assault battalions for the rapid deployment of flexible water pipes between the offensive forces to the newly constructed reservoirs on the west bank of the SC. Water reservoirs made of rubberized tarpaulin (a strong, water resistant canvas) were set up along the length of the pipe. However, despite the geographic intelligence, the IDF intelligence establishment had no inkling of the existence of these engineering assault battalions before the War that hindered the targeting of this dynamic infrastructure.

Egyptian Army Training for the Crossing

By 1972, the Egyptian Army was engaged in military and engineering activity on an unprecedented scale. The identification of such activity was only partly identified and understood by IDF intelligence. The training took place in a demo-training area of the SC in the northwestern Nile Delta that was beyond IAF aerial photo coverage. IDF SIGINT picked up maneuvers of crossing a canal but did not connect these finds with an Egyptian plan to cross the SC. These maneuvers included assembling diverse types of bridges, laying down the rest of the fording equipment that was to be used for breaking through the embankment on the Bar-Lev Line and laying water pipes. Ground forces advanced toward the bridges, crossed them in special combat groups, and deployed for the expected counter-attack.

On the artificial island of El Ballah in the northern SC that was created to enable the two-way passage of ships, Egyptian Army engineers held a series of exercises in which they sprayed water on sand embankments with hoses connected to high-powered pumps (Shazly 1987). Performed directly in front of the Israeli lookout posts, this information was transmitted to S1 who, with the help of geologists, determined that the hosing was intended for opening vehicle passages in the Bar-Lev embankments. However, this interpretation was not accepted by the other intelligence sections nor by the management. The disagreement led to overheated discussions, and no report presenting these findings was disseminated.

During the 1973 crossing offensive, the hosing technique was used successfully upon sandy embankments in the northern and central sections of the SC but was ineffective on the southern and more consolidated silt-clay embankments (Shazly 1987).

Egyptian War Preparations on the Home Front

The Egyptian home front also prepared infrastructure for the coming war. This included amassing emergency stockpiles and protecting strategic facilities from IAF attacks. During the War of Attrition, Egypt began constructing defensive arrangements at strategic targets, such as protective walls and heavily strengthened bunkers at oil reservoirs, and electrical transformers.

In 1972 additional steps taken to protect the home front from attack were identified by S1 in oblique imagery. These included:

- Construction of ten large oil storage depots, most of them underground and well-camouflaged.
- Construction of reservoirs in the Delta and Nile Valley.
- Covering bridges and dams on the Nile with protective nets and deploying barrage balloon units against low-flying aircraft at the bridges.

- Paving access roads and building jetties near important bridges on the Nile. This infrastructure was designed to enable floating bridges to be thrown across the Nile in case the existing bridges were damaged.
- Constructing fake surface-to-air missile bases to mock C2 strongholds of the highest level of Egyptian command.

Discussing and Disseminating Geographic Intelligence

The First Four Spatial Finds Lead to an Organizational Feud

After the S1 discovery of the parallel-pair assault revetments and their realization that the Egyptian Army was working toward a large-scale crossing operation, it was decided to circulate a departmental intelligence review that summarized the general aspects of the crossing along with the perspectives of each of the SIGINT- and HUMINT-based research intelligence sections (S6 and S7). These sections did not, however, provide input, explaining that they didn't have the time to summarize the situation. S1, unwilling to wait, decided that the review would deal only with the ground aspect of the crossing and that the other intelligence sections could contribute their information on other aspects later. In November 1971, 3 years after Kain's initial discoveries, a confidential S1 intelligence review was disseminated within the intelligence entitled "Ground Preparations for the Canal Crossing." Their conclusion was that the Egyptian Army was preparing a large-scale crossing along the entire length of the SC and was protecting the crossing equipment to limit possible damage to it and the crossing forces.

The review forced the research sections to examine the topic. However, S6 completely ignored S1's revelations. One of the reasons may have been that S6 had access to sources that S1 was not privy to which led them to assume S1 was being preemptive. During this period, S1 staff were also barred from attending higher echelon intelligence meetings with the superpowers and foreign military intelligence services. S7, on the other hand, dealt with S1's conclusions in a business-like way, stating that construction of floating bridges across the SC would take a rather long time. They concluded that while the Egyptians were assembling pontoons and dropping them in the water, the IDF would have time to mobilize its entire reserve workforce.

Kain refused to accept S7's estimate of the time needed to assemble the bridges, so he began studying material on water obstacle crossings during World War II. After an exhausting series of conflicts with S7 staff, he convinced them to recalibrate their assessment of the time needed to put the bridges in place. Finally, in early 1973, S7 disseminated a confidential report on the crossing locations along the SC, and affiliation with Egyptian Army units, and the time required for setting up bridges. The report was highly technical but as such ignored the unaccepted fact that the Egyptian Army had restructured and was fully trained for carrying out the SC crossing assignment.

The Debate over Egyptian Readiness for a Canal Crossing

Approximately 3 months before the 1973 War, the head of S1 called an internal meeting to discuss "if the Egyptian army has completed ground preparations for a canal crossing" (Dekel-Dolitzky 2010). The motivation for the meeting was the discovery that the Egyptians had begun building earthen embankments along the access roads leading to the SC. S1 assumed that their purpose was to prevent IDF ground forces from observing movement on the roads. The construction work began south of Little Bitter Lake (Figs. 10.1, and 10.7) after several months of quiet during which the Egyptians had made no improvements in ground preparations along the SC.

Due to the perceived importance of the issue within S1, all the officers and NCOs in the S1 subsections were invited to the meeting, even those who didn't deal with the subject of the crossing. This was unusual, especially as a cease-fire was in effect at the time and no one took much interest in the geographical research. Surprisingly, a representative of Section 6 attended, although the head of main intelligence research was absent. The S6 representative insisted only issues relating to ground preparations be discussed and refused to discuss the Egyptian Army's readiness for war. S6 thought that the road embankment project would eventually extend to all the SC crossing sectors, indicating that a military confrontation would probably not mature for many months.

S1 claimed that the Egyptian Army had basically completed ground preparations for crossing the SC and that the road concealment embankments were unnecessary. A meeting summary was not disseminated probably due to the disagreement.

Egyptian Army Preparations for Offensive Crossing of the Suez Canal

In early October 1973, the Egyptian Army declared a state of alert under the guise of a combined-forces exercise. IDF intelligence went on alert and began tracking Egyptian military activity but decided it was only an exercise. Weather conditions prevented photo reconnaissance flights, so the only incoming information was from SIGINT and ground observations. The latter reported construction enhancing the descents to the water. On October 2, a S1 internal intelligence report on irregular activity for improving the descents along the west bank of the SC to the water was circulated.

On October 4, an air reconnaissance sortie was carried out. From the imagery obtained, S1 rapidly identified large-scale activity and improvements in the descents to the water. Sandbags blocking the descents had been removed, and vehicles with crossing equipment had taken cover in the assault revetments, as S1 had suggested would happen several years before. The rapid and immediate though superficial aerial photo scan and interpretations indicated that physical improvements encompassed nearly 80% of the descents to the water; furthermore, removal of the sandbags from the descents was not a normal occurrence. Due to the potency of the finds, this information was immediately conveyed to higher echelons and the chief of staff.

The spatial determinations were then reexamined toward preparation of an extensive geographic intelligence report. However, the reexamination showed that only 50% of the descents were improved but emphasized that such a rapid and wide-scale activity could mean only one thing according to S1 – an offensive SC crossing was imminent!

The change in the reported percentage of descents ignited internal tension that distracted commanders from focusing on the significance of the finds. The draft of the report was sent for verification to the head of main intelligence research (Fig. 10.4). The head severely reprimanded the head of S1 since he had already reported to the chief of staff the initial find that 80% of the descents to the water had undergone enhancement and now he reads in the report draft that only 50% of the descents had been improved!! This inconsistency of S1 was rapidly conveyed within the intelligence units leading to a loss of credibility of S1 within the intelligence agencies. The report, though, was disseminated but now included an explanatory paragraph by S6 stating that their interpretation of these data was that the Egyptian preparations were part of a combined-forces exercise; S1's assessment that linked this intensified large-scale field activity with war was expunged from the published report.

On the following day at 1400, October 6, Egypt launched their planned war offensive. They easily crossed the SC with five infantry divisions (despite fierce resistance of the small amount of unprepared and surprised IDF forces within the strongholds of the Bar-Lev Line) and captured ground extending up to 10 km east of the SC. The performance of the Egyptian Army was remarkably like the model traced and explained by S1 that began in 1968.

Discussion and Conclusions

The Egyptian October 1973 offensive crossing of the Suez Canal unfortunately verified the geographic intelligence of S1. S1's tactical observations were ground truthed by the IDF troops who captured the west bank of the SC at the end of the war and from captured documentation. Sa'ad al-din Shazly, the Egyptian Army chief of staff at the time, later published a document stating that Soviet-supported planning for the 1973 attack began in 1968 and considerably predated the regrowth of Egyptian operational capability (Shazly 1987). The sequence of events that he described is remarkably like the geographic intelligence produced by Benny Kain and his team.

This case highlights how single researchers have the potential to make a difference. However, geographic intelligence units and their personnel, perceived as a demographic and cognitive minority in relation to larger SIGINT and HUMINT practices, is often constrained with respect to convincing parallel and higher echelons of the significance of GEOINT assessments. The military hierarchic environment further strains the struggle for GEOINT researchers to stand their ground and emphasizes the prime importance of reporting accurate observations. This situation further emphasizes the comprehensive organizational dynamics measures intelligence agencies must facilitate to be relevant and provide a holistic and relevant intelligence picture.

The lessons learned from this study remain relevant for current GEOINT agencies in several ways: (1) The current proliferation and ease of access of imagery combined with the use of geographic information systems provides powerful tools for geospatial analysis. Geospatial intelligence, however, should be studied in coordination with other intelligence sources. (2) Whereas visualization and the high recurrence video-like imagery is very easy to obtain, these technologies often portray GEOINT as an online visualization and imagery service and do not provide necessary resources for establishing terrain intelligence and research that originates from imagery and its derived spatial and temporal datasets. (3) Online satellite imagery can be used as a complementary tool for spatial research. Despite being partly constrained by resolution, recurrence, and processing capabilities, it opens a new window for open-access and public-oriented imagery intelligence that can possibly lead to public pressure on administrations. It surely can be used to identify large infrastructure such as many of the features described in this paper.

This paper exemplifies how geographical intelligence can independently generate strategic intelligence and should never be ignored. Tracking and analyzing the extensive Egyptian military preparations for crossing the Suez Canal in tactical and high-resolution detail is a sterling example of how geographic intelligence research based upon visual data exceeded other collection and SIGINT- and HUMINT-based research in the IDF. The professional human factor also stands out as a key tool behind geographic intelligence research, namely, gathering and deciphering evidence followed by original thinking and a deep understanding of military and civilian doctrine and the terrain.

The positive side of the 1968–1973 drama is that following Egypt's success in crossing the Suez Canal, and despite their subsequent embarrassing military defeat by the IDF, Egypt transmitted the 1973 War as a military success. This sense of superiority enabled and possibly led Egyptian President Anwar Sadat to be the first Arab leader to initiate and achieve a peace agreement with Israel, despite the strict uncompromising and asymmetrical demand of land for peace (Oren 2017).

Acknowledgments This paper is partly based on a chapter in Dekel-Dolitzky (2010). It dedicated to the late Major (reserves) Benny Kain who prior to his dismission from long and exceptional reserve duties was also the first to foresee that toward the turn of the century, the underground and tunnels will be the emerging terrain application for terror and guerilla warfare against Israel.

References

Bar-Joseph, U. (2012). *Watchman fell asleep, The surprise of Yom Kippur and its sources*. State University of New York Press, Albany, N.Y., U.S.A.

Barrowman, R. E. (2007). *Geospatial intelligence: The new intelligence discipline*. Norfolk: United States Joint Forces Command.

Bar-Siman-Tov, Y. (1988). The Bar-Lev line revisited. *The Journal of Strategic Studies, 11*(2), 149–176.

Brugioni, D. A. (1996). The art and science of photoreconnaissance. *Scientific American, 274*(3), 78–85.

Clark, R. M. (2013). Perspectives on intelligence collection. The intelligencer. *Journal of US Intelligence Studies, 20*(2), 47–45.

Colwell, R. N. (1985). *Manual of photographic interpretation*. American Society of Photogrammetry. Falls Church, VA, USA

Dekel-Dolitzky, E. (2007). The intelligence battle concerning the Suez Canal crossing – independent terrain research anticipated the crossing year earlier. The center for intelligence heritage – Malam, 50:34–37 (in Hebrew).

Dekel-Dolitzky, E. (2010). *Groundless intelligence – Success and failure in IDF geographic intelligence*. Tel Aviv: Ely Dekel Books, 310 pp. (in Hebrew).

Hoffmeier, J. K., & Moshier, S. O. (2006). New paleo-environmental evidence from North Sinai to complement Manfred Bietak's map of the eastern Delta and some historical implications. *Timelines: Studies in Honour of Manfred Bietak, 2*, 167–176.

Hoskins, H. L. (1944). The Suez canal and the outlook for Egypt. *American Political Science Review, 38*(1), 110–119.

Karabell, Z. (2003). *Parting the desert: The creation of the Suez Canal*. New York: Knopf.

Oren, M. B. (2017). *Six days of war: June 1967 and the making of the modern Middle East*. A Presidio Press book. The Random House Ballantine Publishing Group: New York.

Oren, A., & Roskin, J. (2015). State security and geography – A theoretical framework of ten main connections. The 11th International Conference of Military Geosciences. Annapolis. Abstract.

Roskin, J. (2018). The role of terrain and terrain analysis on military operations along the turn of the 20th–21st Millennia – A case study of selected IDF battles – this issue.

Roskin, J., & Tsoar, T. (2017). Late quaternary chronologies of the northern Sinai-northwestern Negev dunefield and their palaeoclimatic and palaeoenvironmental implications. In Y. Enzel, & O. Bar-Yosef (Eds.), *Quaternary environments, climate change, and humans in the levant* (pp. 493–504). https://doi.org/10.1017/9781316106754.056. Cambridge.

Roskin, J., Katra, I., & Blumberg, D. G. (2014). Particle-size fractionation of eolian sand along the Sinai–Negev erg of Egypt and Israel. *Bulletin, 126*(1-2), 47–65.

Sanchez, A. (2009). Leveraging Geospatial Intelligence (GEOINT) in Mission Command. Army Command and General Staff Coll Fort Leavenworth KS School of Advanced Military Studies.

Shazly, S. (1987) *Crossing the Canal* (trans: Hebrew). Tel Aviv: Ma'arachot Publishers.

Shlaim, A. (1976). Failures in national intelligence estimates: The case of the Yom Kippur War. *World Politics, 28*(3), 348–380.

Chapter 11
The Role of Terrain and Terrain Analysis on Military Operations in the Late Twentieth to Early Twenty-First Century: A Case Study of Selected IDF Battles

Joel Roskin

Introduction

The paper evaluates the impact of terrain on past battles in an effort to predict the importance of terrain in future western military engagements. Where necessary, army forces use terrain for positional advantage (U.S. Army Field-Manual-100-5_operations 1993). They maneuver in terrain to bring fire power to the enemy and use firepower against the enemy to cover their movement. Up to the late twentieth century, there were officers and military publications that accredited weather and terrain as affecting offensive and defensive military action more than any other physical factor (equipment, weapons, supplies) (Peltier and Pearcy 1966; U.S. Army Field-Manual-100-5_operations 1993; Hatheway and Stevens 1998; Winters 2001; Carter and Veale 2013). However, the impact of terrain and terrain analysis on past combat outcomes has often not been studied adequately (Dekel-Dolitzky 2010), particularly the relative impact of terrain versus other factors such as communications, firepower and firepower accuracy, leadership, motivation, discipline, etc.

The form of engagement of western armies has changed to a degree since the end of the twentieth century from large-scale movement and engagement in natural surroundings to clashes with guerrillas in urban agricultural and often arid Moslem regions (Stewart 2014), strategically aimed at limiting offensive guerrilla and terror capabilities in western countries. As technology has advanced exponentially (i.e., cyber warfare, UAV reconnaissance, precision targeting systems, satellite communications, etc.), there are decision-makers and western army commanders who

J. Roskin (✉)
Analytical Laboratory, Excavations, Survey and Research Department, Israel Antiquities Authority, Jerusalem, Israel

Geomorphology and Portable Luminescence Laboratory, Leon Recanati Institute for Maritime Studies, University of Haifa, Mt. Carmel, Israel
e-mail: yoelr@bgu.ac.il

P. L. Guth (ed.), *Military Geoscience*, Advances in Military Geosciences,
https://doi.org/10.1007/978-3-030-32173-4_11

maintain that the relative importance of terrain and military geosciences for present and future warfare has declined (Burmaoglu and Sarıtas 2017).

Nevertheless, even though technological developments have somewhat reduced the relative importance of terrain for combat outcomes there is currently less emphasis on the importance of terrain, I still wish to argue that terrain is often important in combat between guerilla and conventional forces and soon may become even more important.

The paper uses the term *terrain* to refer to the physical setting of the battles studied. Terrain underpins military engagement (Doyle and Bennet 2002). In military terms, terrain is an area of ground whether large or small whose physical features affect its suitability for military purposes (Parry 1984).

Methodology

The study systematically reviewed eight offensive operations and battles fought by the Israel Defense Forces (IDF) since 1967 in Egypt, Jordan, Lebanon, Gaza Strip, and the Palestinian Authority territories, which are situated topographically in central Israel: Judea and Samara (Fig. 11.1). The period in which the studied battles occurred coincides with the rapidly growing change in role of technology in warfare. The fact that Israel is in the central Middle East is highly relevant for western armies since the region is a key anticipated offensive intervention site for US and western forces against guerrilla units and conventional forces (McDonald 2009). Studies suggest that insights from recent IDF experience can yield operational, tactical, and technological lessons for larger western armies (Johnson 2014; Cohen et al. 2017).

Because of Israel's diverse physical and political geography, military action takes place in different geomorphic settings that influence military activity in many ways (Fig. 11.1). The specific location of Israel on the Sinai-Israel subplate is very important, and due to its location at the eastern edge of the Mediterranean Sea, it has a varied geomorphic structure and a climate ranging from Mediterranean to arid and hyper-arid. Israel borders on five countries (and two entities: Hamas in the Gaza Strip and the Palestinian territories in enclaves in Judea and Samaria) most of which were drawn by European colonial empires in the early twentieth century without consideration of their obvious natural boundaries. These operation settings have given Israel a robust understanding of the impact of different surface terrains on warfare, particularly the relationship between terrain, mobility, and concealment.

The IDF operations in this study generally involved a brigade-size force. The basic task force in special and larger operations is normally brigade to battalion size (Table 11.1). Both types of operations—special and larger—were initiated primarily as a political response to Arab terrorism or terrorism threats against Israeli civilians (other than Battle 1, Table 11.1). While the study does examine engagement with conventional armies, it mostly analyzes engagement with guerillas (small units

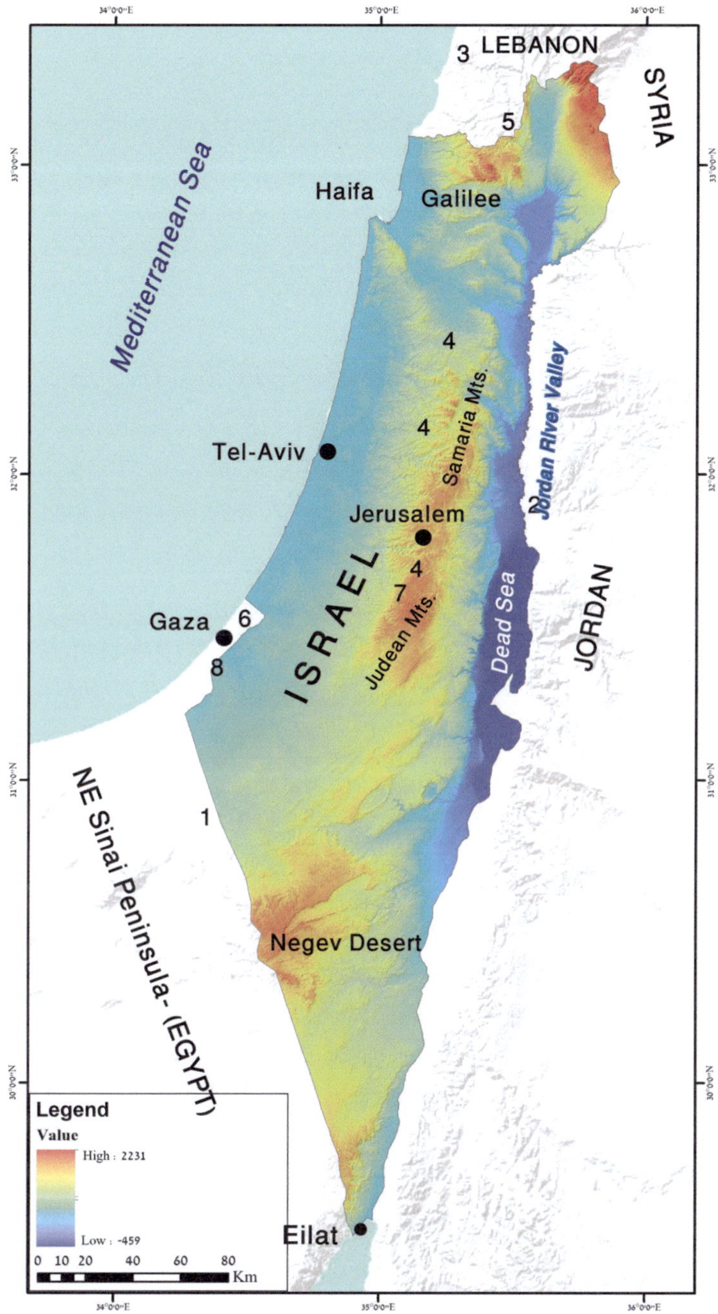

Fig. 11.1 General location of battle sites along the different borders and geographic regions of Israel. See Table 11.1 for battle numbering, coordinates, and details

Table 11.1 Description and analysis of selected IDF battles in chronological order. Note that fatalities are often not a major tool for assessing mission success. See Fig. 11.1 for location of battles

Battle/operation number name, scope, and date	Background: IDF operational and tactical goal	Force size	Location: latitude/longitude	Geomorphology and climate	Terrain assessment	Battle results	Interpretations and tactical and strategic consequences	Comments	Sources
1. Wadi Haradein crossing 8/6/1967 6-Day War	To achieve surprise and control of central Sinai's north-south axis in order to preempt a possible Egyptian Command level counterattack and complete the IDF's command level offensive	Tank brigade	30.910912°/34.185105°Wadi Haradein in northeast Sinai dunefield	Active parallel <20 m high devegetated, loose, active, and steep-sloped (25–30^0) linear dunes. Interdune corridors upon hard calcareous palaeosols and low-energy fine-grained fluvial deposits (LFFDs)/playas/ and loose sand sheets are sporadically blocked by sand deposits and dunes. Arid Mediterranean climate	Understood by IDF intelligence to be assessed unpassable by Egyptians. Trafficable axis in interdune corridors was pre-planned by IDF intelligence VISINT hdq.	Brigade secretly transferred 50 km in 9 hours to carry out surprise attack and destruction of Egyptian tank brigade	Classic example of implementing pre-battle terrain analysis. Field application led to full achievement of tactical and operational goals	This is one of the few positive IDF heritage stories regarding terrain analysis	Tsoar (1995), Dekel-Dolitzky (2010), and Roskin et al. (2011, 2013, 2014)

2. Battle of Karameh War of Attrition battle with the Palestinian Liberation Organization (PLO) and Jordanian Armed Forces 21/3/68	To prevent a PLO terror attack on Israel's 20th Independence Day celebrations. Destruction of PLO bases	Joint infantry-armored brigade-size force vs combined forces of the PLO and the Jordanian Armed Forces (JAF)	31.906867°/ 35.589818° Southern Jordan Valley, Dead Sea Rift, Jordan	Saline clay loam agricultural soils irrigated by open channels and aqueducts Arid climate	Despite months of intelligence and operational preparation did not consider local agricultural practices	27 IDF ATVs and tanks sunk in fields irrigated on the eve of the battle were abandoned. 33 IDF causalities. 150–200 terrorists killed. PLO terrorist head Y. Arafat escaped	Tactical failure but strategic successes: 1. No PLO terrorist attacks on Israel Independence Day 2. Eventually led to Jordanian expulsion of the PLO	One of the main IDF heritage stories of negative terrain analysis The Command General stated in his summary of the operation: "There were a lot of channels and deep and treacherous mud that I never experienced ever before in all the (many) battle zones with the Arabs. Just about every vehicle sank during the operation"	Sutcliffe (1973); http://www.himush.co.il/?section=50; http://scholar.googleusercontent.com/scholar?q=cache:oiZcCbhu56YJ:scholar.google.com/+karameh+jordan&hl=iw&as_sdt=0,5&as_ylo=1960&as_yhi=1980

(continued)

Table 11.1 (continued)

Battle/operation number name, scope, and date	Background: IDF operational and tactical goal	Force size	Location: latitude/longitude	Geomorphology and climate	Terrain assessment	Battle results	Interpretations and tactical and strategic consequences	Comments	Sources
3. Operation “Blue and Brown” 8-9/12/1988	Special operation in retaliation for guerilla attack on IDF base. Aimed at surprising Ahmed Gibril’s guerrilla and terrorist base before dawn	Joint battalion-size force of infantry and commandos	33.752713°/ 35.480781° Lebanese Mediterranean coast south of Beirut by Naame. Thickly vegetated (maquis) wadi with boulders Moist Mediterranean climate	Thickly vegetated (maquis) wadi with boulders Mediterranean climate	Very difficult (slow) trafficability for infantry. Prevented force reaching planned positions before dawn. Shrub thorniness and height was under-calculated. Water pipe obstacle was not accounted for	Majority of the forces did not reach positions before dawn. Force commander’s request to withdraw was declined. The surprise attack was weak. Guerilla retaliation was intense. IDF force commander was killed and 30 dead guerillas	4 IDF soldiers stranded amidst the enemy. When soldiers were found missing, the whole IDF was put on red alert to reinvade Lebanon Soldiers eventually evacuated by helicopter	Exemplifies how disregarding conservative terrain evaluation can lead to serious and strategic consequences	https://he.wikipedia.org/wiki/%D7%9E%D7%91%D7%A6%D7%A2_%D7%9B%D7%97%D7%95%D7%9C_%D7%95%D7%97%D7%95%D7%9D Kheir et al. (2008)

4. Defensive Shield Following a series of Palestinian suicide attacks on Israeli citizens 29/3-10/5/2002	Response to a wave of suicide bombings by Palestinian factions claiming the lives of hundreds of Israeli civilians Two major battles took place in Nablus and Jenin	Several brigades	5 cities of the PA in Northern Judea (Ramallah 31.904961°/ 35.204173°) and Samaria	Urban Palestinian setting usually lacks vegetation cover	Heavily armored bulldozers led the attacking forces and destroyed DEDs along the roads	30 IDF soldiers and 497 Palestinians were killed An immediate significant in the number of suicide attacks on Israelis	IDF demonstrated superiority during the battles, probably due to the massed offensive force and the low soldiership and organizational level of the PA fighters	Urban terrain analysis doctrine for guerrilla warfare was lacking due to the IDF being oriented to a peace-keeping army. Such a doctrine is a complex matter	https://en.wikipedia.org/wiki/Urban_warfare
5. Wadi Saluki battle Second Lebanon War 11-14/08/2006	Offensive crossing of ravine to take control of territory before the cease-fire	Tank brigade with infantry battalion for securing ground control	33.239507°/ 35.475777° Wadi Saluki, southern Lebanon	250 m-deep wadi incised within steep untrafficable limestone and chalk slopes covered by maquis Mediterranean climate	11 out of 24 Merkava tanks descending on dirt road to the wadi. Tanks were hit by Kornet missiles fired from various ground control positions	12 IDF soldiers killed. Unclear if Hezbollah fighters were killed IDF mission failed	Either (a) basic lack of understanding of the impact of terrain by IDF commanders or (b) the decision-makers who ordered the offensive did not consider the terrain. The IDF had no terrain analysis due to its unpreparedness for offensive operations	Higher echelons dismissed a more trafficable alternative to crossing Wadi Saluki because for some reason, it was defined as a "command fire zone." This battle summed up the Lebanon operation which was seen by the Israeli public as a failure	https://en.wikipedia.org/wiki/Operation_Change_of_Direction_11 Harel and Issacharoff (2008)

(continued)

Table 11.1 (continued)

Battle/operation number name, scope, and date	Background: IDF operational and tactical goal	Force size	Location: latitude/longitude	Geomorphology and climate	Terrain assessment	Battle results	Interpretations and tactical and strategic consequences	Comments	Sources
6. Operation Cast Lead 27/12/2008 - 18/01/2009	To prevent/limit intensifying terrorist rocket and mortar fire against Israeli towns and villages despite full Israeli evacuation from the Gaza Strip in 2005	Several infantry and armor brigades vs newly established Hamas guerrilla-style infantry brigades	31.499802°/34.503718°Northern Gaza Strip around the metropolis of Gaza	Intensely urbanized area surrounded by coastal sand sheets and low dunes and agricultural loess and sandy soils in east Semiarid Mediterranean climate (annual rainfall 350 mm)	Ground offensive was delayed by10 days due to IDF central geology unit's evaluation that intense rains softened loess soils impairing trafficability	Gaza City was isolated with "only" several friendly fire IDF casualties	Operational impact of terrain analysis was conveyed to the public	The operation was publicly perceived (in Israel) as successful	http://glz.co.il/newsArticle.aspx?newsid=60841 Dan et al. (1976), Bergman et al. (2014), and Roskin (2016)
7. Brother's Keeper 12/6–30/6/2014	To find 3 kidnapped teens by Hamas terrorists	Several infantry brigades and special units	31.590762°/35.062834°Southern Judea and City of Hebron controlled by PA	500–1000 m-high terraced hard carbonate mountains, steep and, terraced wadis, villages	Free ground movement by foot	Operation failed until hiking guides volunteered to analyze terrain leading to the location of the teenagers' bodies within 24 hours. The bodies were found on a mountainous ridge on farmland owned by one of the kidnappers' relatives	All intelligence agencies seemed quite ignorant regarding tactical terrain		https://en.wikipedia.org/wiki/2014_kidnapping_and_murder_of_Israeli_teenagers

8. Operation Protective Edge 8/7 – 26/8/2014	To prevent/ limit intensifying terrorist rocket and mortar fire on Israeli towns and destroy offensive cross-border tunnels	Several infantry and armor brigades vs established Hamas guerrilla-style infantry brigades	31.258514°/ 34.315495°Gaza Strip: 1–3 kilometers along the armistice line with Israel	Agricultural loess and sandy loess soils Semiarid Mediterranean climate	Summer aridity did not constrain soil trafficability Tunnel detection and destruction from air and by engineering units was very tedious and slow	32 offensive tunnels were destroyed. Despite thousands of sorties, the rocket fire rates from urban areas and underground facilities did not slow down	A lack of comprehensive terrain analysis relating to cross-border tunnel and shaft detection, and a lack of protocols for tunnel destruction, impaired IDF combat effectivity Lack of combined terrain-intelligence analysis impaired detection of urban and rocket underground facilities	Israel's defense minister publicly stated that it would only take a few days to destroy the offensive tunnels. The operation was publicly perceived (in Israel) as a non-victory	https://he.wikipedia.org/wiki/%D7%9E%D7%91%D7%A6%D7%A2_%D7%A6%D7%95%D7%A7_%D7%90%D7%99%D7%AA%D7%9F

against conventional military units). The battles often began with cross-border ground movement by Israeli forces (except for Battles 3 and 5).

The study describes each battle with reference to its date, background, operational and tactical goal, force size, location, geomorphology and climate, terrain assessment, outcomes, and interpretations (Table 11.1). Battle data are based on the personal experience of the author and were retrieved from open military, internet, and scientific sources with the aim of analyzing the effects of terrain and terrain evaluation on battle outcomes.

Results

Positive Impacts of Terrain Analysis on Battles

The ground battles analyzed in Table 11.1 were influenced in different ways by the terrain on which they were fought. The quantitative (fatalities) and perceived outcomes of the analyzed battles range between tactical and strategic failures and successes. Some battles involved tactical failures that did not impede the strategic successes of the engagement (i.e., Battles 4 and 8 in Table 11.1) and vice versa. The battles exemplify how specific physiographic features affect offensive and defensive operations. The various types of operations against guerilla groups in different settings demonstrate the different effects that terrain and urban environments have on military action.

When battles were predicated by accurate terrain evaluations and the recommendations based on these evaluations implemented at brigade level or lower, the outcome was perceived as successful (i.e., Battles 1 and 6, Table 11.1). However, in the remaining six battles, the terrain assessment was either incomplete or only partly followed through, contributing to a non-decisive battle outcome.

Terrain is important for the IDF operations analyzed here due to their cross-border offensive nature: that the defending force had the defensive advantage of knowing the terrain while the IDF was often under political pressure to achieve rapid goals (Kaplan et al. 2005; Harel and Issacharoff 2008). In some cases, as in Battles 3–5, time pressure may have caused the IDF to pay insufficient attention to the tactical implications of the terrain. In Battle 3, IDF decision-makers defined the goal, ground targets, timing, and size and composition of the force. When briefed by the terrain analysts that given the size and nature of the force, the entire force would not be able to transverse the route toward the target offensive before dawn, it was too late to make changes (N.M., personal comm.).

Terrain Analysis Is Not Always Produced

The study findings and interpretations imply that the IDF has surprisingly not carried out consistent, systematic, in-depth terrain evaluations in recent decades. Terrain evaluation greatly assisted the Wadi Haradein mission in 1967 and serves as an outstanding example (Battle 1, Table 11.1) (Dekel-Dolitzky 2010). In Operation Defensive Shield (Battle 4), no terrain analysis was carried out as the political environment was oriented toward peace-keeping and defensive coordination with the Palestinian Authority rather than offensive action (Luft 2002; Kaplan et al. 2005). In contrast, the location of the bodies of three kidnapped youths in 2014 (Table 11.1, Battle 7) was attributed to hiking guides who had experience in basic and uncomputerized terrain analysis who volunteered in the search, this followed 3 weeks of failed searches directed by different intelligence agencies. Also in Operation Protective Edge, the lack of (sophisticated) terrain analysis of tunnels and their environment led the operation to drag on (49 days) way beyond the anticipated length, making it one of Israel's longest operations (Chorev and Shumacher 2014).

Terrain Analysis and Strategic and Political Implications

In the case of the IDF, terrain has greatly affected the outcomes of engagements that often had strategic and political implications (Battles 2, 3, 5, 7, 8, Table 11.1). Historically, as well, in the Middle East, there may also have been a terrain effect in various operations by western armies, such as Operation Eagle Claw (failed rescue of American embassy hostages in Iran in 1980) where they failed to consider the problem of windblown fine-grained sand (Kamps 2006).

Terrain as Force Majeure

The data in Table 11.1 reveal a complex relationship between armies, terrain, and terrain evaluation, and other factors may come into play and dictate operation outcomes, regardless of the terrain assessment. This complexity may partly explain why military mistakes linked to terrain are often accepted by top army commanders as force majeure (Harel and Issacharoff 2008) and are downplayed when lessons are learned. Take, for example, the statement by the IDF General Command regarding Battle 2, Table 11.1. From the battle reports, there is no evidence that terrain analysts were reprimanded for misreading the agricultural and weekly/diurnal irrigation schedule in the battle zone (see Terrill 2001).

Discussion

Terrain Trends in Guerrilla Warfare Engagement

In defensive situations, the guerilla forces' detailed knowledge of terrain seems one of their few remaining advantages in the face of technologically superior western forces. Guerillas have traditionally positioned fortified defensive positions in very steep mountainous regions, forests (Siroky and Dzutsati 2015), or urban settings (Cohen et al. 2017). These landscapes greatly offset the technological and firepower of western forces.

The guerillas forces around Israel have recently improved their defensive and offensive capabilities. They fortify their positions using subterranean features such as caves and tunnels, which provide good concealment from UAVs and communications intelligence. These positions also protect rocket firing pads. These recent changes have dramatically improved guerrilla survivability. Underground infrastructure has been developed by Hamas in the Gaza Strip (Roskin 2015) and Hezbollah in Lebanon following unilateral Israeli withdrawal from both areas (Johnson 2014; Cohen et al. 2017).

The research suggests that despite the tremendous technological superiority of western militaries like the IDF, the tendency of guerrillas to combine underground infrastructures with traditional military-like infrastructures represents an important challenge that should be addressed. The decrease in IDF terrain analysis since the end of the twentieth century combined with the underground efforts of the Arab militias constrains the IDF in attempting to wage effective tactical combat on the ground. Operation Protective Edge exemplifies this point.

Current Considerations for Terrain Analysis

The above reasons partly explain why concealed fortifications are a growing operational challenge to western military forces and their terrain analysis communities (Shay 2014; Marcus 2017; Richemond-Barak 2017; Tallis et al. 2017). Military technology indeed provides sophisticated high-resolution spatial and temporal imaging equipment and push-button terrain analysis (Perez et al. 2017). However, small-scale, low-signature land utilizations such as tunnels require experienced specialists with a rare and essential combination of technical, field, geomorphic-scientific, and military expertise combined with intelligence from other sources (Hatheway and Stevens 1998). Relevant and effective terrain analyses can only be achieved by a combination of supreme terrain technology and experienced manpower. For example, remote thermal-sensing analysis in combat environments requires an understanding of natural and human physical, atmospheric, and subsurface air temperature properties and physical diurnal thermal properties and their effect on enemy materials, maneuverability, and concealment (Sobrino et al. 2016).

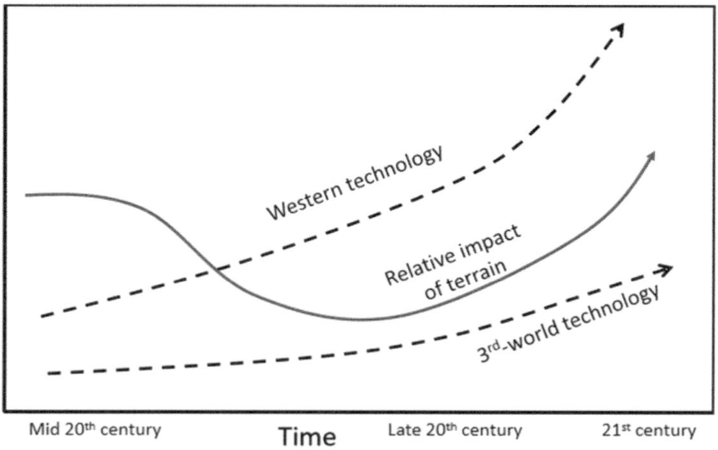

Fig. 11.2 Proposed conceptual model explaining how the importance of terrain and terrain analysis has changed in the context of conflicts between western militaries using sophisticated military technologies and third world entities

In a Conceptual Model

A conceptual model is proposed here (Fig. 11.2) which summarizes the changes in the importance of terrain and terrain analysis with reference to the dynamics of technology and the proliferation of conflicts between third world entities and sophisticated western militaries. The left side of Fig. 11.2 relates to the mid-twentieth century when technologies were relatively less developed, and terrain was relatively important. Since then, however, technology has improved exponentially, and the technological gap between western armies and guerillas has widened commensurately, giving the west increasing and relative technological advantage. The result is that guerillas now seek to limit their exposure to western military remote detection and fire technologies and exploit the advantages that terrain (e.g., urban terrain and subterranean terrain) offers for concealment. As a result, terrain has again emerged as an important factor in modern engagements.

Conclusions and Outlook

This short study presents a selected time series of the impacts of terrain and terrain evaluation on eight engagements fought by the Israeli Army mainly against guerrilla forces. Based on open-source data, it clearly demonstrates that terrain and the application of terrain analysis significantly affect battle outcomes. Nowadays, military operations are still strongly affected by terrain and the accurate/inaccurate terrain analysis provided to field commanders. However, the relative importance of the nature and use of terrain has evolved over time and transformed battle scenarios, for example, the growing use of subterranean space by Arab guerillas. Further study is required to semiquantitatively identify the scenarios and scales at which terrain has significant relative impact on combat.

Terrain evaluation is important for many of the combat scenarios now facing western forces in deprived regions. Since terrain is generally and relatively stable, most terrain analyses, especially those used by the higher command echelons, can be established well before battle and then updated prior to execution. Therefore, to some degree, the relative impact of terrain on combat should be fully implemented prior to combat. In scenarios that seem to have a low relative terrain impact, terrain evaluation is still required as planned courses often change and require additional input to complete the mission. Evaluation of terrain impact on the foe as well as the foe's quality of terrain analysis should also not be ignored. However, effective high-resolution terrain analysis is complex and resource consuming.

To conclude, we cannot ignore the prospect that sometime soon our planet may again experience major clashes, possibly initiated by emerging superpowers, that could remind us of past military lessons and the role of terrain in achieving operational and tactical military goals.

Acknowledgments Peter Guth is warmly thanked for his advice and editorial comments on the manuscript. N.M. (confidential) is thanked for sharing with me recollections on past terrain analysis. Two anonymous reviewers are thanked for their detailed review. Ruth Friedman is appreciated for editing. Semion Polinov (U. of Haifa) helped prepare the map of Israel.

References

Bergman, N., Sholker, O., Roskin, J., & Greenbaum, N. (2014). The Nahal Oz Reservoir dam-break flood: Geomorphic impact on a small ephemeral loess-channel in the semi-arid Negev Desert, Israel. *Geomorphology, 210*, 83–19.

Burmaoglu, S., & Sarıtas, O. (2017). Changing characteristics of warfare and the future of Military R&D. *Technological Forecasting and Social Change, 116*, 151–161.

Carter, T. A., & Veale, D. J. (2013). Weather, terrain and warfare: Coalition fatalities in Afghanistan. *Conflict Management and Peace Science, 30*(3), 220–239.

Chorev, H., & Shumacher, Y. (2014). The road to operation protective edge: Gaps in strategic perception. *Israel Journal of Foreign Affairs, 8*(3), 9–24.

Cohen, R. S., Johnson, D. E., Thaler, D. E., Allen, B., Bartels, E. M., Cahill, J., & Efron, S. (2017). From cast lead to protective edge. Lessons from Israel's wars in Gaza. Rand Corporation www.rand.org. 226 pp.

Dan, J., Raz, Z., & Koyumdjiski, H. (1976). *The soil associations of Israel: Bet Dagan*. Israel: Volcani Inst. (in Hebrew).

Dekel-Dolitzky, E. (2010). *Groundless intelligence – Success and failure in IDF Geographic military intelligence.* Ely Dekel Books., POB 459, Elkana, 44814, 310 pp. (in Hebrew).

Doyle, P., & Bennett, M. R. (2002). Terrain in military history: An introduction. In: *Fields of battle: terrain in military history*, (Vol. 64). Springer Science & Business Media, Netherlands.

Harel, A., & Issacharoff, A. (2008). *34 Days: Israel, Hezbollah and the War in Lebanon*. New York: Palgrave Macmillan. xi + 288 pp. ISBN 9780230604001.

Hatheway, A. W., & Stevens, M. M. (1998). Military geology should be upgraded as the US Army stands down. *Reviews in Engineering Geology, 13*, 233–238.

Johnson, D. E. (2014). *Military capabilities for hybrid war: Insights from the Israel defense forces in Lebanon and Gaza*. Santa Monica: Rand Corporation.

Kamps, C. T. (2006). Operation eagle claw: The Iran hostage rescue mission. *Air Sp Power J, 18*(3), 1–21.

Kaplan, E. H., Mintz, A., Mishal, S., & Samban, C. (2005). What happened to suicide bombings in Israel? Insights from a terror stock model. *Studies in Conflict & Terrorism, 28*(3), 225–235.

Kheir, R. B., Chorowicz, J., Abdallah, C., & Dhont, D. (2008). Soil and bedrock distribution estimated from gully form and frequency: A GIS-based decision-tree model for Lebanon. *Geomorphology, 93*(3), 482–492.

Luft, G. (2002). The Palestinian H-Bomb: Terror's winning strategy. *Foreign Affairs, 81*(4), 2–7.

Marcus, R. D. (2017). Learning 'Under Fire': Israel's improvised military adaptation to Hamas tunnel warfare. *Journal of Strategic Studies, 42*(3), 1–27.

McDonald, E. (2009). Military Geosciences for the 21th century. US military research initiatives in the DRI. 20 pages.

Parry, J. T. (1984). Terrain evaluation, military purposes. In C. W. Finkl Jr. (Ed.), *The encyclopedia of applied geology* (pp. 570–580). New York: Van Nostrand Reinhold Company.

Peltier, L. C., & Pearcy, G. E. (1966). *Military geography* (Vol. 30). Princeton: Van Nostrand Company.

Perez, D., Banerjee, D., Kwan, C., Dao, M., Shen, Y., Koperski, K., Marchisio, G., & Li, J. (2017). Deep learning for effective detection of excavated soil related to illegal tunnel activities. In IEEE Ubiquitous Computing, Electronics and Mobile Communication Conf.

Richemond-Barak, D. (2017). *Underground Warfare*. New York: Oxford University Press.

Roskin, J. (2015). Geographic influences on the development of clandestine smuggling tunnels between Sinai (Egypt) and Rafah (Gaza Strip) and Terror-Guerilla tunnels between the Gaza Strip and Israel. 11th International Conference on Military Geosciences, Annapolis. Presentation.

Roskin, J. (2016). Analysis of recurring sinking events of armored tracked vehicles in the Israeli agricultural periphery of the Gaza Strip. In *Military geosciences and desert warfare* (pp. 339–352). New York: Springer.

Roskin, J., Porat, N., Tsoar, H., Blumberg, D. G., & Zander, A. M. (2011). Age, origin and climatic controls on vegetated linear dunes in the northwestern Negev Desert (Israel). *Quaternary Science Reviews, 30*(13), 1649–1674.

Roskin, J., Katra, I., Porat, N., & Zilberman, E. (2013). Evolution of Middle to Late Pleistocene sandy calcareous paleosols underlying the northwestern Negev Desert Dunefield (Israel). *Palaeogeography, Palaeoclimatology, Palaeoecology, 387*, 134–152.

Roskin, J., Katra, I., & Blumberg, D. G. (2014). Particle-size fractionation of eolian sand along the Sinai–Negev erg of Egypt and Israel. *Geological Society of America Bulletin, 126*(1–2), 47–65.

Shay, S. (2014). Syria and the sub terrain warfare–the Hamas connection. In Updating Israel's National Security Doctrine, Policy Paper, The Herzliya Conference.

Siroky, D., & Dzutsati, V. (2015). The empire strikes back: Ethnicity, terrain, and indiscriminate violence in counterinsurgencies∗. *Social Science Quarterly, 96*(3), 807–829.

Sobrino, J. A., Del Frate, F., Drusch, M., Jiménez-Muñoz, J. C., Manunta, P., & Regan, A. (2016). Review of thermal infrared applications and requirements for future high-resolution sensors. *IEEE Transactions on Geoscience and Remote Sensing, 54*(5), 2963–2972.

Stewart, A. K. (2014). US Army agriculture development teams: A grassroots effort in Afghanistan supporting development and tackling insurgency. *Science and Diplomacy, 3*(1), 70–87.

Sutcliffe, C. R. (1973). The East Ghor Canal project: A case study of refugee resettlement, 1961–1966. *Middle East Journal, 27*(4), 471–482.

Tallis, J., Bauer, R., & Frey, L. (2017). ISIL'S Battlefield tactics and the implications for homeland security and preparedness. *Journal of Terrorism Research, 8*(3), 24–42.

Terrill, W. A. (2001). The political mythology of the Battle of Karameh. *The Middle East Journal, 55*, 91–111.
Tsoar, H. (1995). Desertification in northern Sinai in the eighteenth century. *Climatic Change, 29*(4), 429–438.
U.S. Army Field-Manual-100-5_operations, 1993 Headquarters 100-5 Department of the Army, Washington, DC, 1 1993. 163 pp.
Winters, H. A. (2001). *Battling the elements: Weather and terrain in the conduct of war*. Baltimore: JHU Press.

Chapter 12
Passive Seismic Survey of Sediment Thickness, Dasht-e-Nawar Basin, Eastern Afghanistan

Thomas J. Mack

Introduction

Determining the thickness of aquifers is an important component of assessing the groundwater resources of a region. Airborne and satellite data have proved to be highly useful for helping to remotely determine the type or nature of surface sediments and rocks in Afghanistan. However, determining the thickness of unconsolidated aquifers remains a critical question that can only be addressed by on-site geophysical surveys or drilling. Most geophysical surveys for assessment of deep sediments typically require survey lines of thousands of meters, and both geophysical surveys and drilling operations are time-consuming, therefore subjecting field crews to increased security risks. In Afghanistan, traditional geophysical surveys and drilling operations are severely limited by security constraints.

As part of an investigation of mineral resources in Afghanistan, the Department of Defense Task Force for Business and Stability Operations (TFBSO) consulted with the US Geological Survey (USGS) on the exploration of minerals at more than 24 locations around the country (Peters et al. 2007, 2011). One of the areas of interest identified for further investigation was the Dasht-e-Nawar Basin (DN Basin) (Fig. 12.1) in east-central Afghanistan. The DN Basin is a closed basin where the occurrences of pegmatites, mineralized springs, dry lake sediments, and surface brines indicated that it could be a favorable host for lithium resources (Stillings et al. 2015). Extractable lithium deposits generally occur as brine in areas of thick sediments in such evaporative basins.

T. J. Mack (✉)
U.S. Geological Survey, Pembroke, NH, USA
e-mail: tjmack@usgs.gov

P. L. Guth (ed.), *Military Geoscience*, Advances in Military Geosciences,
https://doi.org/10.1007/978-3-030-32173-4_12

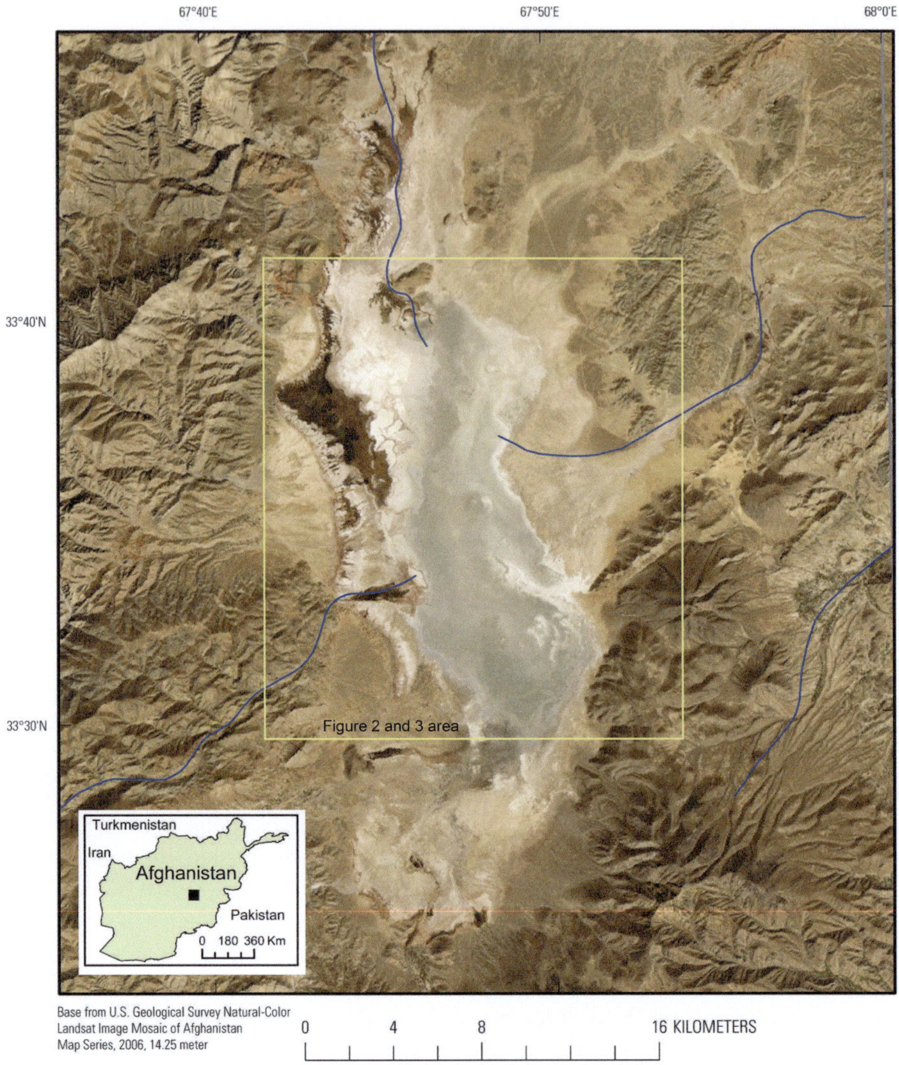

Fig. 12.1 Dasht-e-Nawar Basin, east-central Afghanistan

Between June and September 2014, the TFBSO contracted with a geological exploration company, Centar America,[1] to conduct a geophysical, drilling, and core and groundwater sampling exploration program in the basin. Exploration drilling is an expensive and time-consuming operation requiring a skilled field crew and logistical support. It was important to minimize the drilling time, particularly with

[1] Any use of trade, firm, or product names is for descriptive purposes only and does not imply endorsement by the US Government.

fieldwork in hazardous areas, while optimizing the drilling sites to areas where there is the greatest possibility of encountering thick sediments. Therefore, DN Basin geophysical surveys were performed to identify areas of thick basin sediments as drilling targets. The geophysical surveys used were a gravity survey, conducted by New Era Geophysics for Centar, and a passive seismic survey coordinated and interpreted by the USGS, with field assistance from Centar. The passive seismic survey had not previously been used in Afghanistan; the application of this technique and comparisons to the results of the gravity survey were the emphases of the study described in this paper.

Methods and Equipment

This study compared the results of a passive seismic survey against an independent traditional gravity anomaly survey. The gravity survey used a Scintrex CG5 semiautomated gravimeter with survey locations collected at intervals of 250 m on approximately east-west lines (e.g., A02 to D02 and B03 to E03, Fig. 12.2). The lines were spaced approximately 1 km apart, and 770 readings were collected over a period of 20 days (New Era Geophysics 2014). A large number of measurements were collected along the east-west lines for delineation of gravity differences. Interpretation of the gravity data also required a precise elevation map and measurements of soil and rock specific density. Sediment thickness was interpreted by New Era Geophysics (2014) from residual gravity in milligal (mGal) using a crustal density of 2.65 g per cubic centimeter (g/cc) and a sediment density of 1.85 g/cc. Assuming crustal and sediment densities are homogenous in the area, the residual gravity map should mimic sediment thickness. Digital terrain elevation data were acquired by New Era Geophysics with a differential GPS survey conducted prior to the gravity survey. Soil and rock samples were collected by Centar, and measurements of density were determined by the USGS (Stillings et al. 2015). The gravity survey was interpreted by New Era Geophysics, at several east-west lines crossing areas of greatest residual gravity and at one north-south line. The gravity surveys were used in this investigation as provided by New Era Geophysics.

The passive seismic survey (Stillings et al. 2015) used a portable (handheld) broadband seismometer, the Tromino (Moho 2017). This equipment weighs approximately 1 kg and is about 10 cm^3 in size. Each point measurement requires 20–30 minutes in the field and is operated by one person. The technique is termed passive in that unlike other seismic techniques, it records ambient noise without an external sound source. Passive seismic interpretation is made on a point basis and does not require additional data, other than an understanding of the geologic setting of the investigation site. Although relatively new for assessment of sediment thickness, several investigations have found this method to be capable of estimating sediment thicknesses, in the 10's to 1000 meter range (Ibs-von Seht and Wohlenberg 1999; Delgado et al. 2000; Parolai et al. 2002; Lane et al. 2008; Haefner et al. 2011).

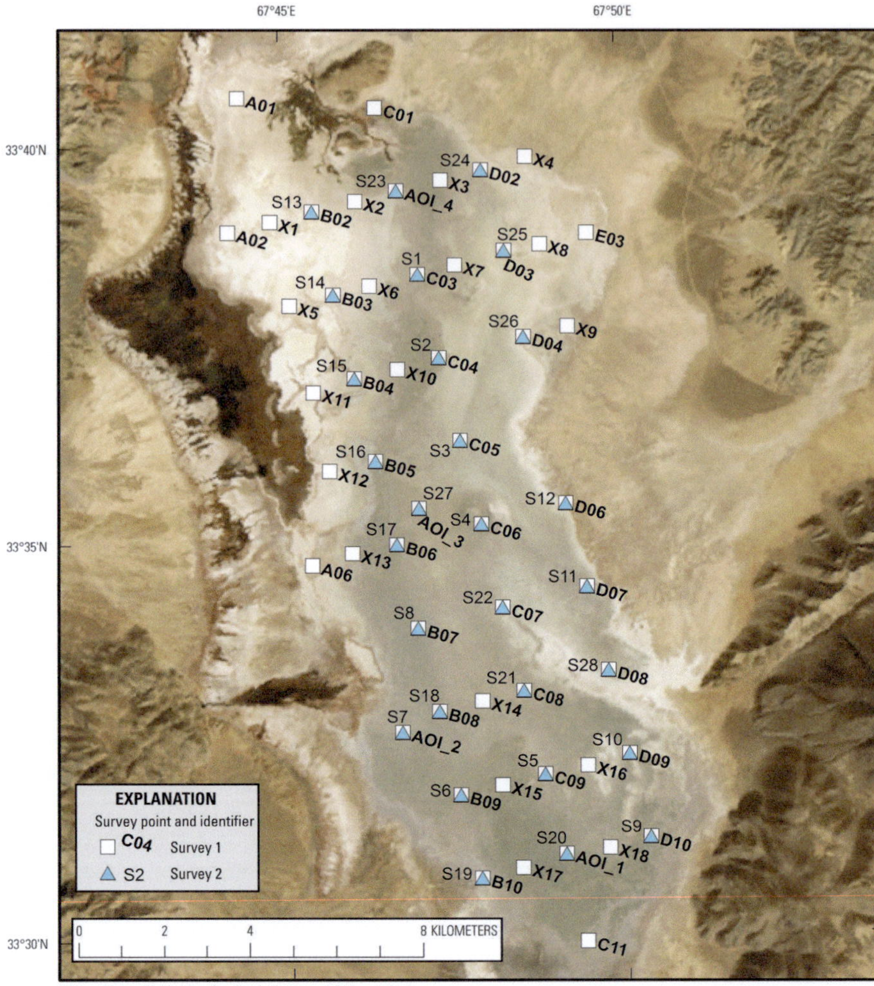

Base from U.S. Geological Survey Natural-Color Landsat Image Mosaic of Afghanistan Map Series, 2006, 14.25 meter

Fig. 12.2 Survey locations and identifiers in the Dasht-e-Nawar Basin, Afghanistan

Passive seismic investigation of sediment thickness is often termed the H/V method because by this method, the spectral ratio of the horizontal (H) and vertical (V) components of the seismic noise is used to determine the site resonance frequency (SESAME 2004). Resonance frequency, (f_{r0}), is a logarithmic function of sediment thickness (Z) following the equation:

$$Z = a f_{r0}^{\ b} \tag{12.1}$$

where the coefficients *a* and *b* are determined empirically for a site or region. Seismic data were processed using Grilla version 6.4 software (MoHo 2017) following guidelines given by SESAME (2004). There were no depth to bedrock measurements in the DN basin for local calibration of the sediment thickness function; therefore, coefficients derived by Lane et al. (2008) ($a = 94$, $b = -1.324$) from two contrasting settings were used for this investigation. Since confirmation of sediment thickness from borings was not available, locally derived coefficients were generated from borehole minimum thickness estimates as a check of the viability of the published coefficients for this setting. Any potential error associated with this assumption is not known. Interpretation was performed on 20-second trace windows, with triangular window smoothing, using the GRILLA software to provide an average peak resonance frequency for calculation of sediment thickness using Eq. 12.1. The resonance peak over time was also examined to get an understanding of the range of the frequency peaks that may indicate subsurface topography at a measurement point.

The H/V method requires a contrast in acoustic impedance, a product of material density and seismic velocity, equal to or greater than 2 to 1 of the bedrock and overlying sediments (Lane et al. 2008). This method becomes ineffective with a poor contrast in acoustic impedance, as may occur with unconsolidated sediment overlying semi-lithified sediments or weathered bedrock. Prior to the passive seismic survey, it was not known whether the bedrock underlying the DN Basin would provide a sufficient acoustic contrast necessary for sediment thickness interpretation.

Due to security concerns in areas surrounding the investigation site, USGS personnel could not travel to the site. Instead, the two individuals who conducted each survey were given brief data collection instructions before the measurements were collected by Centar (Terrance Cameron, CENTAR American Technical Services, written commun. 2014). One individual collected the first set of measurements, and a second individual collected the second set of measurements. In the first survey, from 29 July to 2 August, 50 measurements were collected for 20 minutes at each station (Fig. 12.2). In the second survey, from 20 to 22 September, 28 measurements were collected for 30 minutes at select stations, at either repeated survey locations from survey 1 or at additional stations (Fig. 12.2).

Results

Figure 12.3 shows a color-contoured map of the residual gravity provided by New Era Geophysics (2014). Figure 12.3 also shows the interpreted sediment thickness (New Era Geophysics 2014) at the deep regions along a north-south centerline (C01 through C11) of the basin. The northern lobe of the basin was interpreted to have sediment thicknesses of about 120 m near points C03 and C04. The area in the center of the basin between C05 and C06 is interpreted to have sediment thicknesses of about 50 to 80 m. The southern deep lobe had the greatest interpreted sediment thicknesses of 145 m near C09 and 160 m near AOI_1.

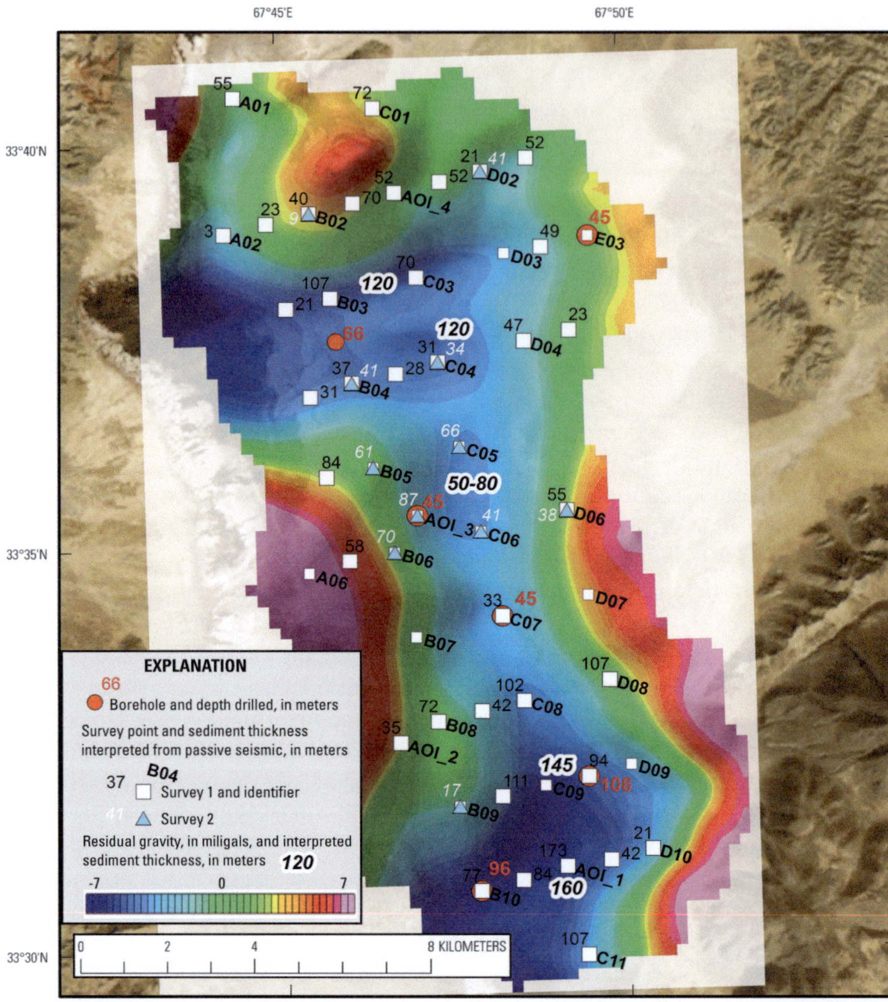

Base from U.S. Geological Survey Natural-Color Landsat Image Mosaic of Afghanistan Map Series, 2006, 14.25 meter

Gravity survey from New Era Geophysics (2014)

Fig. 12.3 Residual gravity image and interpreted sediment thickness and passive seismic interpreted sediment thickness in the Dasht-e-Nawar Basin, Afghanistan. (Survey 1 sites X1–X15 are not labeled to improve map readability; see Fig. 12.2)

Results of the two passive seismic surveys are shown in Fig. 12.3 overlying the residual gravity image for comparison. The DN passive seismic survey data are provided in Stillings et al. (2015). The seismic surveys also indicate a northern and southern deep lobe of the basin with sediments greater than 100-m thick (survey locations B03, C09, and AOI_1, Fig. 12.3). The maximum interpreted sediment thickness, using literature-derived coefficients, was 107 and 173 m in the northern and southern basin lobes, respectively. Due to site conditions, and the fact that the operators had limited training and no experience, about 80 percent of measurements were poor, and some measurements were not interpretable (Stillings et al. 2015). However, most interpreted sediment thicknesses were generally consistent with surrounding or coincident measurements. Overall, the seismic survey illustrates a reasonable sediment thickness pattern, with thinner sediments at the margins of the basin and thick sediments toward the center, similar to the gravity survey. However, the seismic and gravity surveys differed considerably in the area of C04. In this area, the gravity survey indicated a maximum sediment thickness of 120 m, and, in contrast, the two seismic measurements indicated a sediment thickness of 31 and 34 m. Nearby seismic measurements at B04 and D04 and the point between these measurements (X10, Fig. 12.2) indicate similar sediment thicknesses. The area along the points B04, C04, to D04 may be an area where the gravity survey is not accurate, possibly due to unmapped lithology changes. However, some of the passive seismic measurements, although interpretable, were found to be of poor quality. In this case, additional seismic measurements, with experienced field operators, would likely resolve the basin thickness.

A wide range in frequency resonance peaks, for example, as shown in Fig. 12.4, may result in a large confidence interval (Stillings et al. 2015). However, this does not necessarily reflect the quality of the measurement. A wide range in frequency resonance peaks, and corresponding large confidence interval, may occur with a response from a sloping target surface, or one with considerable buried topography. For example, the resonance frequency measured at site B03 (Fig. 12.3) has an average peak centered at 0.91 Hz (Fig. 12.4) and a broad region from about 0.8–2 Hz, indicated by the dark red response in Fig. 12.4, that may represent shallower and

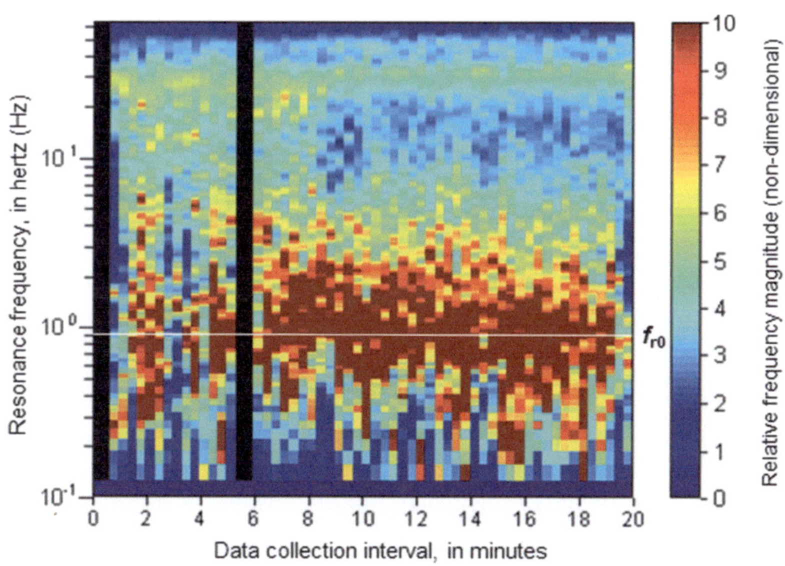

Fig. 12.4 Resonance frequency over the passive seismic collection interval at site B03 illustrating the range in resonance peak, Dasht-e-Nawar Basin, Afghanistan

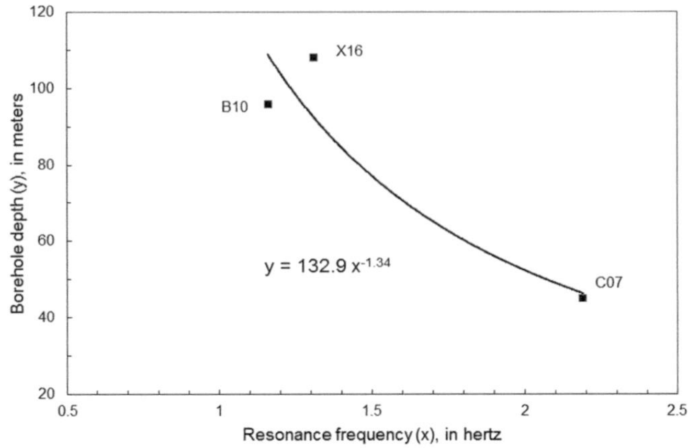

Fig. 12.5 Estimated sediment thickness function based on borehole depths in the Dasht-e-Nawar Basin, Afghanistan

deeper surfaces. One interpretation of the broad resonance peak is that the site overlies a small pocket of thick sediments surrounded by thinner sediments.

Six boreholes were drilled in the DN Basin, and unfortunately none of the boreholes encountered bedrock to provide a definitive check on the geophysical survey interpretations. However, three of the boreholes, located at C07, B10, and X16 (between C09 and D09, Figs. 12.2 and 12.3), were drilled deeper than the sediment thickness interpreted from the seismic surveys. Based on these borehole depths, the gravity-estimated thicknesses also appear to be underestimated. If the boreholes are close to the bedrock surface at these points, then adjusted sediment thickness function coefficients for Eq. 12.1 for the basin, based on the data and function shown in Fig. 12.5, are $a = 132.9$ and $b = -1.34$. Using these coefficients, the estimated sediment thickness would be deeper than both the gravity and passive seismic survey results using published coefficients. The reinterpreted, passive seismic, maximum thicknesses would be at least 150 m deep in the northern area of the basin and 247 m in the southern area. These results indicate that more field calibration data are needed to accurately calculate sediment thickness in the DN Basin.

Discussion

The passive seismic surveys at Dasht-e-Nawar were negatively affected by conditions including poor ground coupling and strong winds. About 74 percent of the first seismic survey was interpretable, but only 43 percent of the second survey was interpretable. The second survey was particularly poor, and discussion with the field operator indicated that the equipment was probably placed in a manner that led to a poor coupling. Data collection issues could likely have been counteracted by protecting the equipment from wind and removing loose surface sediment at measurement points. Although the quality of many measurements was poor, a resonance frequency peak could still be identified, and the interpreted sediment thicknesses were generally similar to those of the independent gravity survey. Additionally, at the few sites where there were two sediment thicknesses interpreted from the seismic surveys (collected on different dates), most of the thicknesses were found to be similar. For example, at site C04 (Fig. 12.3), thicknesses of 31 and 34 m were calculated, and at site B04, thicknesses of 41 and 37 m were calculated. The passive seismic survey was interpreted using literature coefficients. Although the H/V method has been found to be robust in suitable conditions, Haefner et al. (2011) noted the possibility of underestimation of sediment thickness in clays. Sediments in the Dasht-e-Nawar Basin are primarily thick clays (Stillings et al. 2015) which may contribute to the underestimation of sediment thicknesses noted at three points based on the borehole information. However, it appears that both gravity and seismic surveys underestimated sediment thickness in the basin. Future passive seismic surveys, or reanalysis, would benefit from depth to bedrock information for developing better coefficients for Eq. 12.1. Yet even with issues associated with inexperienced operators, the passive seismic technique was found to be a rapid and low-cost alternative to a more time-consuming and data-intensive geophysical survey. This technique could be used as a first-pass screening tool for geophysical surveys or drilling programs, or it may provide calibration points for airborne geophysical surveys in Afghanistan and other areas where rapid or efficient data collection is important.

Acknowledgments The author wishes to acknowledge the field efforts of Michael Chornack (USGS, retired) and the reviews and contributions provided by James Degnan and Rodney Sheets (USGS). The author also wishes to acknowledge the efforts of Terrance Cameron, CENTAR American Technical Services, in providing field assistance and data. This work was funded by the US Department of Defense Task Force for Business and Stability Operations. This paper is number 342 in the USGS Afghan Project product list.

References

Delgado, J., López Casado, C., Giner, J., Estévez, A., Cuenca, A., & Molina, S. (2000). Microtremors as a geophysical exploration tool: Applications and limitations. *Pure and Applied Geophysics, 157*, 1445–1462.

Haefner, R. J., Sheets, R. A., & Andrews, R. E. (2011). Evaluation of the horizontal-to-vertical spectral ratio (HVSR) seismic method to determine sediment thickness in the vicinity of the South Well Field, Franklin County, Ohio. *Ohio Sci J, 110*(4), 77–85. http://kb.osu.edu/dspace/handle/1811/52793.

Ibs-von Seht, M., & Wohlenberg, J. (1999). Microtremors measurements used to map thickness of soft soil sediments. *Bulletin of Seismological Society of America, 89*, 250–259.

Lane J. W. Jr, White, E. A., Steele, G. V., & Cannia, J. C. (2008). Estimation of bedrock depth using the horizontal-to-vertical (H/V) ambient-noise seismic method. In: Symposium on the Application of Geophysics to Engineering and Environmental Problems, Philadelphia, Pennsylvania, Proceedings: Denver, Colorado, Environmental and Engineering Geophysical Society, p. 13, 6–10 Apr 2008.

MoHo. (2017). http://moho.world/en/tromino/. Accessed 19 Nov 2017.
New Era Geophysics. (2014). Gravity survey over the Dasht-e-Nawar saline lake, Afghanistan. TFBSO contract 0034-14-C-0050, 5 Sept 2014, Harare, Zimbabwe, p. 18.
Parolai, S., Bormann, P., & Milkert, C. (2002). New relationships between *Vs*, thickness of sediments, and resonance frequency calculated by the H/V ratio of seismic noise for Cologne Area (Germany). *Bulletin of Seismological Society of America, 92*, 2521–2527.
Peters, S. G., Ludington, S. D., Orris, G. J., Sutphin, D. M., Bliss, J. D., & Rytuba, J. J. (2007). Preliminary non-fuel mineral assessment of Afghanistan. U.S. Geological Survey Open-File Report 2007-1214, p. 810.
Peters, S. G., King, T. V. V., Mack, T. J., & Chornack, M. P. (Eds.) (2011). Summaries of important areas for mineral investment and production opportunities of nonfuel minerals in Afghanistan. U.S. Geological Survey Open-File Report 2011-1204, p. 810.
SESAME. (2004). Guidelines for the implementation of the H/V spectral ratio technique on ambient vibrations, measurements, processing and interpretation. Project EVG1-CT-2000-00026. http://www.geopsy.org/index.html. Accessed 19 Dec 2015.
Stillings, L. L., Mack, T. J., Chornack, M. P., Kalaly, S. S., Ahmadi, M. I., & Akbar, A. Q. (2015). A summary of data collected by the U.S. Geological Survey at Dasht-e-Nawar, Afghanistan, in support of lithium exploration, June–September 2014. U.S. Geological Survey Open-File Report 2015–1059, p 70. https://doi.org/10.3133/ofr20151059. Accessed 19 Dec 2015.

Chapter 13
Silver Bullets and the Paradox of Plenty: Natural Resource Development in Afghanistan

Lt Col Drew Craig

Introduction

In 2010 a *New York Times* article based on United States Geological Survey research (Peters et al. 2007) purported that Afghanistan was host to over 1 trillion US dollars of mineral resources. Almost immediately, commentators, governments and even the US military were promoting the mineral resources of Afghanistan as a proverbial silver bullet which was "enough to fundamentally alter the Afghan economy and perhaps the Afghan war itself" (Risen 2010).

However, to any economic geologist, the prospectivity of Afghanistan was no surprise, and indeed there was ample evidence dating from the 1950s to 1970s of the extensive exploration programmes conducted by British, German and Soviet geology teams (Peters et al. 2007). Indeed, historical evidence indicates that lapis lazuli was sent to the pharaohs of Egypt and copper mining goes back as far as 1000 BC (Global Witness 2012).

Afghanistan's geology has all the qualities necessary to make it some of the most highly prospective ground in the world: it is geographically located at a strategic, economic nexus between Europe and the growth markets of Central Asia and Southeast Asia; it is underexplored and unexposed to modern exploration technology; it sits within the economically rich Tethyan Metallogenic Belt (Richards 2015), has a complex geological history and as such holds significant promise for the discovery of a range of commodities and potentially world-class deposits. But a range of issues, the most obvious of which is the past 30 years of prolonged conflict, have made it unattractive for investment when there were lower risk opportunities elsewhere.

L. D. Craig (✉)
SO1 Geology, Royal Engineers, 66 Works Group, 170 Infrastructure Support Group, Royal Engineers, British Army, Nottingham, UK
e-mail: 832craig@armymail.mod.uk

P. L. Guth (ed.), *Military Geoscience*, Advances in Military Geosciences,
https://doi.org/10.1007/978-3-030-32173-4_13

The Paradox of Plenty

The development of a nation's natural resources sector is not without risks as they can in fact lead to instability and even conflict resurgence. Natural resources do not make conflict inevitable, but their presence, especially in low-income and poorly developed countries, can significantly exacerbate the risks of conflict developing, can prolong conflict if it breaks out and can ultimately make it harder to resolve (Ross 2003; Berman et al. 2015). The discovery of abundant natural resources in an emerging market can paradoxically lead to stagnant growth or even economic contraction. This is commonly referred to as the Paradox of Plenty or the Natural Resources Curse.

The paradox occurs as a country begins to focus all of its energies on a single industry, such as mining, and neglects other major sectors. As a result, the nation becomes overly dependent on the price of commodities, and overall gross domestic product becomes extremely volatile. Additionally, government corruption often results when proper resource rights and an income distribution framework are not established in the society, resulting in unfair regulation of the industry.

There are also additional risks to a country's economy and industrial base. For example, Dutch disease occurs when an economic boom, driven by the discovery of natural resources, can lead to a destabilising of currency, a reduction in economic competitiveness and low interest rates. Combined, these factors can lead to an exodus of investment and reduced economic potential.

The mineral resources that appear to cause the most harm to stability are largely oil and hardrock minerals including coltan (columbite-tantalite), diamonds, gold and other gemstones (Craig 2015). Note the hardrock minerals are all low-volume/high-value commodities which makes their transport and sale all the harder to track, control and in turn tax by the authorities. The ability for most of these commodities to be produced artisanally makes them ideal as revenue generators for insurgency. Of the 15 intrastate conflicts in Fig. 13.1, 14 are related at least in part to mineral and hydrocarbon wealth.

The Paradox in Recent History

One commentator (The Economist, 2010), upon reading the 2010 news of Afghanistan's mineralogical endowment, was quick to draw analogies with the Democratic Republic of the Congo (DRC) where minerals are very much seen as a curse and a catalyst perpetuating the conflict and unrest that has infested Africa's second largest and most mineral-rich country for over 30 years (Ross 2003). UN estimates for the mineral wealth of the DRC exceed USD$ 24 trillion (UN 2011).

At present, the DRC remains a semi-stable state with large tracts of land under government control, mostly in the west. However eastern DRC is awash with a

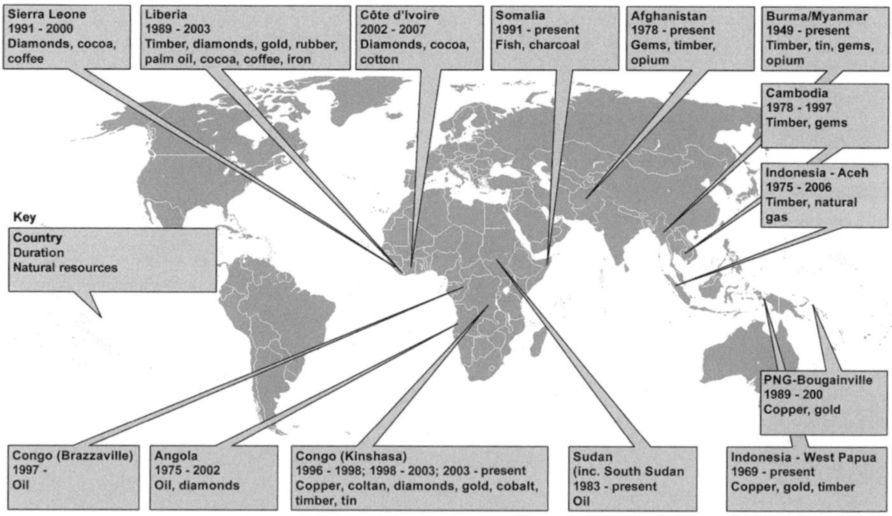

Fig. 13.1 Intrastate conflicts with links to natural resources in the last 60 years. (Adapted and updated from Ross (2003))

variety of different rebel groups, originating from within the DRC and from the neighbouring countries of Rwanda, Uganda and the Central African Republic, which are taking advantage of the lack of a strong rural state presence to seize control of the area's mineral riches (Global Witness 2009). As a result of the persistent instability, the Kinshasa government has not been able to capitalise on the wealth to deliver the improvements in infrastructure, wealth dissemination and local economy that would undoubtedly have be realised in less unstable jurisdictions.

The DRC is therefore a prime example of where the presence of world-class mineral resources continues to act as a destabilising element. It should be noted however that the exploitation of copper and cobalt has occurred for over 90 years in the southeastern Katanga province, thereby indicating that high levels of risk may be tolerated for the chance to exploit potential world-class deposits. Indeed Craig (2015) illustrated that several of the major producing countries, for both gold and copper, are jurisdictions that would be regarded and measured as being poor in terms of doing business, risk (both security and political) and also corruption. Examples include the Democratic Republic of the Congo, Uzbekistan and China.

Countries that have successfully broken the resources curse and are now reaping the post-peace dividend of mineral and hydrocarbon resources include Angola, Liberia and Sierra Leone. All are seen as success stories where the respective governments, following resource-supported conflicts, have managed to mitigate a degree of the corruption and bad practice that had previously prevented them from utilising their mineral resources as a catalyst to drive infrastructure development, employment and secondary and tertiary industries (Craig 2015).

Foundations for Success

Several initiatives, such as the Kimberley Process and the Extractives Industry Transparency Initiative (EITI), have reduced the risk of natural resources sustaining conflicts or catalysing the onset of hostilities. However, the utilisation of a state's natural resources as a means of stabilising and developing, by international donors and peace-keepers, is still poorly understood in terms of both (1) the fundamentals required by a state to generate wealth through the exploitation of natural resources and (2) the long-term strategy that needs to be followed by donors and their agents to achieve a sustainable solution.

Where an almost immediate impact can be made in stabilisation is through supporting the host government and its agencies, the most obvious being the Ministry of Natural Resources or Mines and also any indigenous Geological Survey. Assisting them to help themselves would in all but a few cases be deemed to be the optimal approach.

Beyond the stabilisation phase in a conflict country, transition into the development phase, in a more permissive environment, may permit a wider range of field-based activities that are less dependent on the presence of either state or private security forces and therefore less expensive. By achieving even a semi-permissive environment in isolated areas, perhaps provincially, the opportunity grows for gaining access to prospective geology and attracting investment.

Early stage exploration can lead to localised economic stimulus with the employment of local labour, purchase of local produce, and support in community engagement projects, such as medical clinics or veterinary visits. Development work can start to improve existing infrastructure in an area, and a mine itself can lead to major infrastructure development.

The one exception to this scenario is that of artisanal mineral production during stabilisation which may provide an economic incentive to divert manpower away from insurgency by providing a viable, commensurate alternative to taking up arms. In addition to economic factors, political, security and social dimensions should be understood, to ensure that the initiative will have a positive impact on stabilisation. Special attention needs to be paid to the risk of socio-economic exclusion and of the economic benefits being monopolised and abused by non-state actors.

Figure 13.2 illustrates a simplified organisational structure for the state management of natural resources and highlights areas where mentoring, training and support might be provided, ultimately geared towards improving the efficiency of the minerals sector.

As seen from Fig. 13.2, the most important requirement for a nation to build a mining industry is the legislation that determines the ownership, consumption, distribution and governance of the natural resources (Evans 2010). This legislation forms the foundation upon which governmental departments and agencies can manage and exploit the resources for the benefit of the country.

Development and exploitation of natural resources are all but impossible without a robust and attractive legal framework and as importantly the organisational

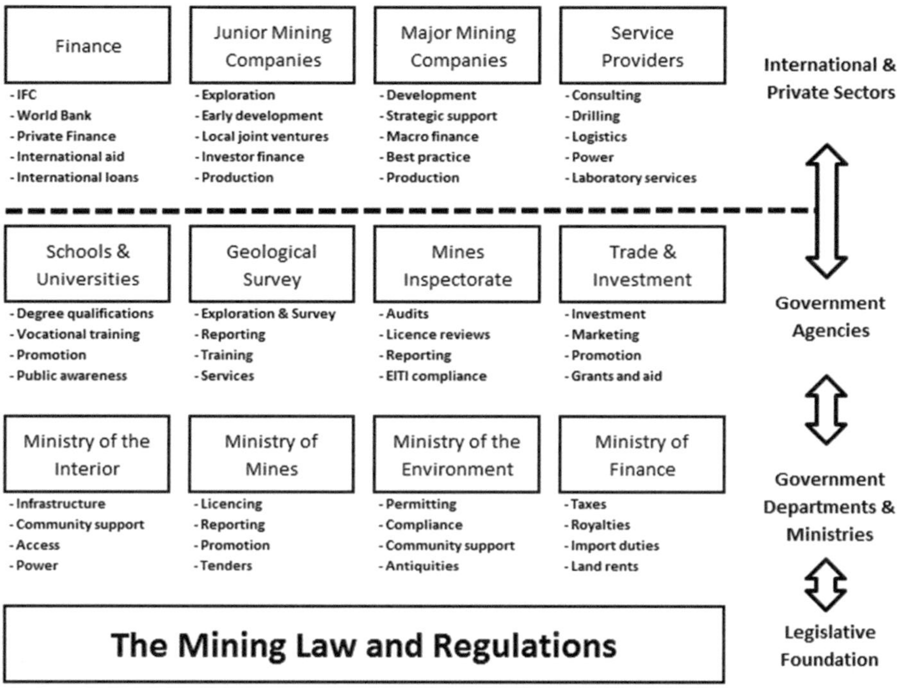

Fig. 13.2 Simplified, generic organisational structure for the state management of natural resources. (Developed from Craig (2015))

architecture to implement and enforce it. The absence of this architecture can lead to unsanctioned exploitation, often artisanal in nature, going unchecked. This might benefit local communities in the short term but will not benefit the country as the product is unlikely to be taxed or provide any royalties. Low-bulk, high-value commodities, such as precious stones or gold, which could provide substantial revenue for the government, are often the main targets for illegal artisanal production and, as noted in earlier sections, can act as a catalyst for destabilisation and further illegal activities.

For investment, one of the most important aspects of any mining law is that of security of tenure. Recent surveys have consistently identified resource nationalisation as being a key risk for resource investment (Ernst and Young 2015). Licence cancellation or non-award may result in protracted and expensive international arbitration which invariably has a devastating effect on investor confidence in a particular country.

Many of the needs illustrated in Fig. 13.2 have been identified by the donor community, and in the next sections, we examine what work has been done to support the vision for Afghanistan's exploitation of its natural resources. We will then shortlist those areas that are still problematic.

Natural Resource Sector Development in Afghanistan

The post-2001 development of natural resources in Afghanistan began in the aftermath of the expulsion of the Taliban from power in 2001–2002. To date, it is estimated from public domain information alone that in excess of USD$ 586 million has been budgeted or spent on the development of the extractives sector in Afghanistan (Table 13.1). These funds have been utilised by more than 20 organisations working on behalf of approximately 8 primary donor organisations to develop both the hydrocarbon and mineral sectors (Table 13.2).

Table 13.1 Estimated donor expenditure on major programmes focusing on the extractives sector in Afghanistan from 2002 to present

Programme	Funding
World Bank – SDNRP 1	USD$ 30 million
World Bank – SDNRP 2	USD$ 52 million
USAID	>USD$ 206 million
TFBSO	>USD$ 282 million
DFID (2015–2018)	USD$ 16 million
Asian Development Bank, AusAID, BMZ, GIZ and others	Unknown
Total	*>USD$ 586 million*

Table 13.2 Examples of the donors and contractor/agents who have contributed to the development of the Afghanistan extractives sector

Donor organisations	The World Bank (WB) Department of International Development, UK (DFID) Department of Defence, US (DoD) Task Force for Business and Stability Operations, US (TFBSO) United States Agency for International Development, US (USAID) Asian Development Bank (ADB) Deutsche Gesellschaft für Internationale Zusammenarbeit, Germany (GIZ) International Finance Corporation (IFC)
Contractor/agent organisations	Deutsche Gesellschaft für Internationale Zusammenarbeit (GIZ) SRK Exploration Services Limited, UK (SRKES) SRK Consulting (US) Limited (SRK) Coffey, UK Flag International, US Mayer Brown, UK DAI, US GAF AG, Germany United States Geological Survey, US (USGS) Adam Smith International, UK (ASI) International Geoscience Services, UK (IGS) ECC Water & Power LLC, US (EWP)

However, despite the efforts of these organisations and their significant funding, progress on the development of both sectors remains intermittent at best, and it will take many years. The Afghanistan Ministry of Mining and Petroleum (MoMP) and the Afghanistan Geological Survey (AGS) still require a series of factors to be favourably resolved before the country might seriously consider itself being on the path to developing a mining industry and in turn have that industry contribute significantly to the country's GDP.

Some of the major programmes are as follows:

United States of America The United States Geological Survey (USGS) operated intermittently in Afghanistan from 2002 to 2015, across a range of initiatives and programmes, notably documenting and publishing on geology and mineral endowment, conducting hyperspectral remote sensing and geophysical surveys and providing training and equipment to the AGS (Peters et al. 2007, 2011; USGS 2015).

From 2010 to 2014, the Task Force for Business and Stability Operations (TFBSO) operated under the direction of the US Department of Defence (DoD). The TFBSO's mission was to help Afghanistan unlock its economic future with focus areas on minerals, energy, indigenous industries, agriculture and information technology. Several of the USGS initiatives benefited from the arrival of the TFBSO as the availability of military security enabled USGS and other consulting subject matter experts to access the field to conduct geological assessments and fieldwork. Key extractives work for the TFBSO included strategic advisory services at government and ministerial level, tender support and international promotion, hydrocarbon infrastructure development and rehabilitation, geophysical surveys, mineral exploration and site evaluation, capacity development and training at the MoMP and AGS and support to artisanal mining activity.

The United States Agency for International Development (USAID) has supported the extractives sector since 2002, through the USGS, and funding for other programmes. It funded the Mining Investment and Development for Afghanistan (MIDAS) programme which was operational from March 2013 to March 2017. The MIDAS programme is focused on three components: legal and regulatory reform, technical assistance to the MoMP and Small and Medium Enterprise (SME) development (USAID 2015). USAID has also funded several hydrocarbon projects, including the Sheberghan Gas Development Project and the Sheberghan Gas Generation Project.

United Kingdom The British Geological Survey (BGS) was commissioned by the British Government's Department for International Development (DFID) in 2004 to provide support to the MoMP and the AGS. Based in Kabul, the BGS team's main focus was in training, generating mineral databases, helping to develop the mineral economy and encouraging good governance (Mitchell and Benham 2008; AGS 2013). Several of the elements were conducted in partnership with the USGS programme. The BGS programme concluded in 2007 due to a withdrawal of funding from the British government.

DFID has also supported the MoMP directly since 2002 through the provision of embedded expertise from Adam Smith International (ASI), a leading independent government advisory consultancy. Their focus has been around supporting the MoMP in the restructuring of the AGS, including an organisational structure review, improvements in departmental management, training plans, capacity building for management and technical staff and the transference of commercial functions to the private sector. This work in ongoing.

In 2015 DFID funded the commercial vehicle of the BGS, International Geological Services (IGS) to support a programme focusing on three areas: environment laboratory facilities at the AGS, materials testing laboratory facilities at the AGS and international promotion and marketing in support of the MoMP. This work is ongoing.

The World Bank The World Bank has supported several phases of work in Afghanistan as part of the Sustainable Development of Natural Resources Projects in Afghanistan (SDNRP) programme. They were also responsible for the tendering of the Aynak copper and Hajigak iron ore projects. Work has either completed by the World Bank directly or through associated consultants such as GAF-AG, a German company with specialisations in earth observation and geoinformation solutions (GAF-AG 2018). GAF-AG projects included the training of mines, establishing an online cadastre and inspectorate system, the Aynak Compliance Monitoring Project and supporting various MoMP and AGS marketing and educational activities such as visits to mines in other countries and facilitating attendance at international mining events. Phase 2 of SDNRP is due to run through to 2017 with other projects either ongoing or in planning, including the Aynak Compliance Monitoring Project, Capacity Building at the AGS (CBAGS) and other project management and monitoring programmes. The World Bank engaged Deutsche Gesellschaft für Internationale Zusammenarbeit (GIZ) to manage and support several of their programmes, in turn supported by contributions from the German Federal Ministry for Economic Cooperation and Development.

There were many more international programmes with cross-cutting sectoral themes, e.g. gender empowerment, infrastructure development, small to medium scale enterprise (SME), service sector development, technical training and expertise, regional cross-border initiatives, etc. These included:

Asian Development Bank (ADB) – supporting natural resources operations, capacity building for aid coordination and management and railway development study

Australian Aid for International Development (AusAID) – sustainable economic and social development initiatives

German BMZ and USGS – hydrological studies in the Kabul basin

Collectively these programmes have cost in excess of USD$ 586 million (see Table 13.1).

Despite the significant resources expended over the past 16 years, the Afghanistan extractives sector is still far from being anywhere close to generating the income anticipated in 2010. In part, this is due to poor expectation management at almost every level of every programme for both Afghans and international donors alike. Perhaps more importantly is that a significant proportion of those funds have gone on poorly conceived and badly managed programmes. It should be acknowledged that this is not a problem exclusively seen in the extractives sector; it has manifested itself throughout the country since combat operations began in 2001. In the words of Heather Barr of the Human Rights Watch, Afghanistan is a "perfect case study of how not to give aid" (Dhillon 2014).

Despite the obvious desire to exploit the utility of natural resources for the greater good of the Afghan state, the level of communication, cooperation and coordination between the donors, their agents and the Afghan stakeholders continues to fall short of what should be possible. The only published reviews on the expenditure on natural resources are by the Special Inspector General for Afghanistan Reconstruction (SIGAR). SIGAR was created by the US Congress in 2008 to provide independent and objective oversight of Afghan reconstruction projects and activities. SIGAR's 2015 and 2016 reports on US activities in the resource space concluded that there were a lack of overall US strategy, a lack of US interagency coordination, a lack of MoMP capacity and a lack of US sustainability planning. These reports concluded that:

> Unless U.S. Agencies Act Soon to Sustain Investments Made, USD$ 488 Million in Funding is at Risk. (SIGAR 2015a)

> Both the efforts of TFBSO and USAID in this area [extractive industries] produced mixed results... Without addressing the above mentioned problems [numerous], it is unlikely that Afghanistan's extractive industries will develop to their full potential. (SIGAR 2016)

Brinkley (2013) notes that:

> Nearly a dozen government auditing agencies have been warning for almost a decade about their concerns for the way USAID, the Defense Department, and other agencies hand over multimillion-dollar construction projects to private, for-profit contractors and then inadequately monitor what they are doing with the money.

Numerous further viewpoints support the conclusion that the collective donor effort has failed to meet with the cultural, economic, administrative and corruption related challenges experienced in Afghanistan (Farzam and Seerat 2016). Nevertheless, the single constant that remains unaffected by donors and politicians of all sides is the prospective geology beneath their feet and that in itself offers some degree of hope.

Plugging the Gap

The following sections identify where additional work must be focused to enable Afghanistan's natural resources to be prepared for extraction:

Strategic Communication, Coordination and Cooperation Long-term success requires a long-term strategy. Donors must put aside individual and conditional policies and work together alongside the Afghanistan government and its agents to follow the strategy plan. The plan should be prepared by the government utilising the expertise and knowledge of the subject matter experts within the donor community. It is essential that the expectations of both the government and donor are kept to realistic objectives and within feasible time lines.

Such a plan would contribute to the reduction in the risks of duplication, financial wastage and the delivery of substandard capacity building and enhance the prospect of improved transparency and accountability.

Licencing The issue of ground for the purposes of exploration should be on a first-come, first-served basis, avoiding lengthy and time-consuming tenders. This will encourage exploration companies to stake ground which would generate land rent income. A licence contractual commitment for annual expenditure, commonly per unit area under licence, will also prevent speculative staking and ensure that exploration work is carried out. This commitment should extend to the submission of annual reports, completed to international standards, providing evidence of the work completed. Failure to completed work or return reports would lead to licences being revoked and opened up to other companies.

Inspectorate and Functioning Cadastre In order to manage licences both administratively and technically, the government needs to improve the MoMP run inspectorate and the mining cadastre. Whilst there have been efforts to train inspectors, they still lack the administrative framework to complete their jobs effectively, in both Kabul and the provincial centres. The current cadastre management system is ineffective and requires an off-the-shelf commercial solution.

Law Despite several revisions in 2005, 2009 and 2014, the current law and regulations are uncompetitive and poorly understood by the government and its agents. A lack of administrative, legal and technical capacity within the MoMP, exacerbated by parliamentary interference, has stalled progress on five major project tenders and led to at least one preferred bidder withdrawing. Issues still remain over the security of tenure for a licence, most notably during the transition from exploration to exploitation.

Economics One of the few variables that is entirely out with the scope of the Afghanistan government or the donor community is that of the global natural resources economy. Over the past 8 years, both the mineral and hydrocarbon sectors have seen a cyclic global downturn in commodity prices, a reduction in access to investment capital and even major mining companies losing billions of dollars in value. Whilst the minerals sector follows a regular boom and bust cycle, few if any of the donor, or indeed the MoMP, were accounting for this within their development strategy. Additionally, during these downturns it is commonly the exploration budgets of juniors and major natural resource companies that are cut back first, and so given that the majority of Afghanistan's upside is in exploration, there is even more pressure from other risk factors on attracting investment. At the time of writing this paper, the industry is still experiencing a prolonged downturn with few if any market commentators prepared to call the bottom of this commodity cycle.

Market Perception, Media and Marketing Planning Widespread access to the media in Afghanistan, through cell phone and the Internet, makes it all the more necessary for the MoMP and the AGS to have detailed, long-term media plans to attract international mining investment and just as importantly help to educate the Afghan people about the mining industry and how it works.

The MoMP must attend international events to actively promote the Afghan natural resources sector and allay the fears and misperceptions of the mining investment community. Fig. 13.3 illustrates some of the key perceptions that need to be addressed, notably "risk" and "security".

Corporate Social Responsibility (CSR) As mining companies are finding across the globe, the community is key to securing the consent to operate from exploration through the exploitation. Various organisations are currently promoting such initiatives, and with ready access to the media, all stakeholders have the opportunity to influence the success of a project. Equally, those with ulterior motives can derail projects by gaining the upper hand through manipulating market perception (see previous section), and so the MoMP must take a proactive stance on ensuring that CSR forms an integral part of both government and corporate strategies.

Fig. 13.3 Wordle derived from comments collected from 26 delegates during a perception survey of the Afghanistan minerals sector. (Data collected by the author at the Prospectors & Developers Association of Canada in Toronto, March 2014)

Education With many of the elder generation of Afghan geologists and engineers reaching or passing the age of retirement, there is an urgent need to invest in the junior generation. The prolonged instability has prevented junior and graduate Afghans from building and expanding upon their university education. Extensive vocational training is needed to provide a cadre of geologists capable of sustaining a growing industry. Failure to provide this cadre will result in high operational costs due to the requirement to bring in costlier expatriate expertise. The MoMP must also ensure that senior generations of geologists and engineers do not languish in posts, thereby preventing younger generations from advancing up the career ladder. This in turn necessitates adequate provisions for retirement and pensions.

The Afghanistan public also needs to be educated as to how the extractives industry works so as to prevent the perception that foreign companies are "stealing" from the nation (Global Witness 2012).

Female Empowerment Given that females make up half of the population, a society is limiting itself if women are excluded from the technical workforce. Improvements to female integration to society in general have been a high donor priority in Afghanistan. There have been major improvements, but further effort is required to ensure that female Afghans are involved at all levels of public service and private enterprise. Education plays an important part of this process.

Anti-corruption and Transparency Initiatives Afghanistan is preparing for membership of the Extractive Industries Transparency Initiative (EITI) which is a global voluntary standard to ensure transparency of payments from natural resources. Further support and mentoring will likely be needed to assist the MoMP, but ultimately, through EITI, the Government of Afghanistan will become more accountable for the revenues received from the exploitation of Afghanistan's natural resources.

Open Market Stimulation The natural resources sector has been proven to stimulate job creation through primary, secondary and tertiary services during all phases, from exploration, to development and then to production. 2010 figures from British Colombia in Canada (PwC 2011) showed that every dollar spent in the mining industry generated USD$ 1.73 comprising direct (USD$ 1.00), indirect (USD$ 0.53) and induced (USD$ 0.20) output impact. The same dollar also resulted in USD$ 0.91 of GDP and USD$ 0.18 of taxes (direct, indirect and induced; federal, provincial and municipal). The same report also demonstrated mining's ability to act as an employment multiplier. Of a total of 45,703 mining industry jobs, 21,112 (46%) were direct jobs (of which 8195 [18%] from operating mines), 16,590 (36%) were indirect jobs and 8001 (18%) were induced jobs.

Infrastructure and Access Whilst some mega-scale projects in the resource space can fund their own infrastructure, there invariably needs to be a strategic infrastructure plan into which small- to medium-scale extractive operation might link into. Infrastructure includes rail, water, road and power; all of which might necessitate the involvement of multiple ministries and multiple provincial and municipal governing bodies. In exceptional circumstances, the presence of resource projects may comprise a critical component of resource corridor infrastructure. International initiatives may also complement and integrate to broaden the reach and value of infrastructure, for example, the TAPI (Tajikistan, Afghanistan, Pakistan, India) pipeline and the China-Pakistan Economic Corridor (CPEC).

However, for these benefits to be realised, there must be an economic base from which to grow the sector. This requires the government to provide incentivisation, potentially including tax breaks and grants and the underlying employment, taxation and business laws to be transparent and easy to administrate.

Summary

Despite the well-resourced efforts of a number of minor and major programmes since the entry of international forces into Afghanistan in 2001, a number of key issues are apparent which require detailed attention for the Afghans to truly realise the potential of their natural resources:

- Reduction of jurisdictional risk, namely, an attractive mining law and regulations, transparency of ownership and inter-ministerial coordination
- The development of an effective cadastre and inspectorate
- Capacity Development and Sustainability within the MoMP and the Afghanistan Geological Survey (AGS)
- Donor coordination, communication and cooperation
- Education (schools, university, AGS and MoMP)
- Marketing (internationally and nationally; investor perception, Afghan public perceptions)

To use a nation's natural resources as a tool towards stabilisation and development, there are several important points to note:

- Collective planning is essential; across nations, SMEs, organisations and agencies (both military and civilian) towards a clear strategy drafted with full support of the host nation.
- A modern and competitive minerals law is essential as foundation to both industrial and artisanal scale mineral exploitation.
- The host nation's minerals strategy must integrate and compliment wider infrastructure and economic planning.
- Natural resource-related initiatives should be developed and managed within the framework of a broader programme of stabilisation and then development activities.
- Broader programmes should support, or at the very least be deconflicted with, the stabilisation strategy, the legitimacy of the host nation and the efforts of the international community.

- Major considerations for planning include relevance, inclusion, coverage, effectiveness, efficiency, sustainability, impact, speed, coordination, visibility, appropriateness, agency, monitoring and evaluation.
- "Do No Harm". The project's indirect negative consequences should be considered thoroughly – Are these acceptable? How might the programme/project fail or destabilise the political settlement in the short and medium term? This principle needs to be applied pragmatically, enabling swift but considered responses (DFID 2010).

Conclusions

For Afghanistan to become a stable country, it needs an economic base that will enable it to finance its own security and development. Afghanistan's natural resources have the greatest potential to provide the country with that economic base.

Several world-class projects will undoubtedly be discovered and subsequently exploited in Afghanistan in the future, contributing significantly to the country's GDP, reducing reliance on donor aid and bringing widespread benefits in infrastructure, employment and public services.

The international donor community is well aware of the need to develop the extractive sector and to date has spent in excess of USD$ 586 million on natural resource related projects. However, a lack of long-term strategy, poor communication, an absence of coordination and an aversion to cooperation have meant that successes of over 15 years of work have been limited.

Ongoing economic challenges within the natural resource sectors compound upon the failures of the donor community and the challenges of the rampant corruption experienced at every level in Afghan society. Ultimately, in the event of such challenges being overcome, the time necessary to explore and develop sites of prospectivity results in the reality that Afghanistan's natural resources will not make a significant contribution to the state's economy for at least 5–10 years.

Acknowledgements A great deal of the thought and commentary in this paper is derived from the 36 months I spent providing support to the TFBSO as a civilian subject matter expert. During this period, I was responsible for delivering a range of services, ranging from field level surveys and technical reporting to tender and contract negotiations as part of government level strategy. I was also involved with the USAID MIDAS programme and through UK military links kept abreast of DFID activity. My final task with the TFBSO was to compile a comprehensive database of natural resources sector development, and it is this task, in concert with other strategic government-level commissions in a range of global jurisdictions, that has afforded me the insight into both the risks and opportunity presented by an underdeveloped but highly prospective geology.

References

AGS. (2013). Donor assistance: BGS. http://www.bgs.ac.uk/AfghanMinerals/assist.htm. Accessed 25 Apr 2018.

Berman, N., Couttenier, M., Rohner, D., & Thoenig, M. (2015). This is mine is mine! How minerals fuel conflicts in Africa. OxCarre Research Paper 141, Department of Economics, Oxford Centre for the Analysis of Resource Rich Economies. http://www.oxcarre.ox.ac.uk/files/OxCarreRP2014141.pdf. Accessed 25 Apr 2019.

Brinkley, J. (2013). Money pit: The monstrous failure of US aid to Afghanistan. *World Affairs., 175*(5), 12–23.

Craig, A. G. (2015). The minerals sector contribution to stabilisation and development in Afghanistan. In E. Hunger, & T. J. Brown (Eds.). Proceedings of the 18th extractive industry geology conference 2014 and technical meeting 2015, EIG Conferences Ltd, pp. 132–145.

DFID. (2010). Working effectively in conflict-affected and fragile situations summary note. https://www.gov.uk/government/uploads/system/uploads/attachment_data/file/67696/summary-note-briefing-papers.pdf. Accessed 25 Apr 2018.

Dhillon, K. (2014). Afghanistan is the big winner in U.S. foreign aid. In *Time*. Available via http://time.com/43836/afghanistan-is-the-big-winner-in-u-s-foreign-aid/. Accessed 25 Apr 2018.

Ernst & Young. (2015). Business risks facing mining and metals, 2014–2015. http://www.ey.com/Publication/vwLUAssets/EY-Business-risks-facing-mining-and-metals-2014%E2%80%932015/$FILE/EY-Business-risks-facing-mining-and-metals-2014%E2%80%932015.pdf. Accessed 25 Apr 2018.

Evans, A. (2010). World development report 2011 – Background paper: Resource scarcity, climate change and the risk of violent conflict. Centre on International Cooperation. New York University. http://siteresources.worldbank.org/EXTWDR2011/Resources/6406082-1283882418764/WDR_Background_Paper_Evans.pdf. Accessed 25 April 2018.

Farzam, R., & Seerat RA. (2016) The Failure of Foreign Aid in Afghanistan. https://thediplomat.com/2016/10/the-failure-of-foreign-aid-in-afghanistan/. Accessed 28 Apr 2018.

GAF-AG. (2018). GAF-AG Home page. http://www.gaf.de. Accessed 25 Apr 2018.

Global Witness. (2009). War and the militarisation of mining in Eastern Congo. https://www.globalwitness.org/sites/default/files/pdfs/report_en_final_0.pdf. Accessed 28 Apr 2018.

Global Witness. (2012). Copper Bottomed? Bolstering the Aynak contract: Afghanistan's first major mining deal. https://www.globalwitness.org/documents/11537/copper%20bottomed.pdf. Accessed 28 Apr 2018.

Mitchell, C., & Benham, A. (2008). Afghanistan: Revival & redevelopment. http://www.bgs.ac.uk/AfghanMinerals/docs/IndustrialMineralsFinalArticle.pdf. Accessed 25 Apr 2018.

Peters, S. G., Ludington, S. D., Orris, G. J., Sutphin, D. M., & Bliss, J. D. (Eds.). (2007). Preliminary non-fuel mineral resource assessment of Afghanistan: US Geological Survey Open-File Report 2007-1214.

Peters, S. G., King, T. V. V., Mack, T. J., & Chornack, M. P. (Eds.), U.S. Geological Survey Afghanistan Mineral Assessment Team. (2011). Summaries of important areas for mineral investment and production opportunities of nonfuel minerals in Afghanistan: U.S. Geological Survey Open-File Report 2011–1204.

PwC. (2011). Economic impact analysis – Mining Association of British Columbia. http://www.mining.bc.ca/sites/default/files/resources/pwcmining-economicimpactanalysis_1.pdf. Accessed 25 Apr 2018.

Richards, J. P. (2015). Tectonic, magmatic, and metallogenic evolution of the Tethyan orogen: From subduction to collision. *Ore Geology Reviews, 70*, 323–345.

Risen, J. (2010). U.S. Identifies vast mineral riches in Afghanistan. http://www.nytimes.com/2010/06/14/world/asia/14minerals.html. Accessed 25 Apr 2018.

Ross, M. (2003). The natural resources curse: How wealth can make you poor. In I. Bannon & P. Collier (Eds.), *Natural resources and violent conflict* (pp. 17–42). Washington: The World Bank.

SIGAR. (2015a). SIGAR 15-55-AR/Afghanistan's extractive industries report. Special Inspector General for Afghanistan Reconstruction. https://www.sigar.mil/pdf/audits/SIGAR-15-55-AR.pdf. Accessed 25 Apr 2018.
SIGAR. (2015b). SIGAR 16-11-AR/Afghanistan's extractive industries report. Special Inspector General for Afghanistan Reconstruction. https://www.sigar.mil/pdf/audits/SIGAR-16-11-AR.pdf. Accessed 28 Apr 2018.
The Economist. (2010). Oh great, another Congo. http://www.economist.com/blogs/democracy-inamerica/2010/06/afghan_minerals. Accessed 12 Apr 2013.
UN. (2011). DR Congo: UN advises prudent use of abundant resources to spur development. http://www.un.org/apps/news/story.asp?NewsID=39986#.UWfre1eGQ4Q. Accessed 25 Apr 2018.
USAID. (2015). Mining Investment and Development for Afghanistan Sustainability (MIDAS) project. https://www.usaid.gov/afghanistan/fact-sheets/mining-investment-and-development-afghanistan-sustainability-midas. Accessed 25 Apr 2018.
USGS. (2015). USGS Projects in Afghanistan. http://afghanistan.cr.usgs.gov/background. Accessed 25 Apr 2018.

Chapter 14
The Need for Geoscience Inputs in Civil Military Planning and Response

M. H. Bulmer

Introduction

Earth in the twenty-first century is marked with societies ravaged by, or at significant risk of, conflicts, disasters, environmental emergencies, and humanitarian catastrophes. The number of armed conflicts around the world in 2015 was 50 up from 41 in 2014. The conflicts in Syria, Iraq, and Afghanistan accounted for 76% of the total battle deaths in 2015 (Trends in Armed Conflict 2016). In the last decades, national and foreign militaries have been involved in emergency and disaster response, reconstruction, and development roles. At times these have occurred as part of warfighting and counterinsurgency operations, peacekeeping, and peace support operations. The twenty-first-century strategic environment is increasingly best described using the concept of coupled human and natural systems (CHANS) (National Research Council 1999; Sheppard and McMaster 2004; Marina et al. 2011). A myriad of ethnic, religious, ideological, and capability drivers create the human systems that interface and interact with the natural system (Fig. 14.1). Satellite data has allowed annual measurements of the percentage of Earth's plant life (natural primary production) and human's need for food, fiber, wood, and fuel. Large urban areas consume greater than 300 times more NPP than the local area produces creating rising imbalances. Human systems are increasingly being shocked by environmental degradation, geological, hydrometeorological, and space weather events (Fig. 14.2). In addition, there is evidence that human-induced climate change is altering the interfaces and interactions that link human to natural subsystems [e.g., Stern (2007), and IPCC (2014)]. This requires a renewed emphasis in civil military response planning on identifying naturally induced drivers of conflict, disasters, and humanitarian catastrophes. Where possible, the aim is to prevent them

M. H. Bulmer (✉)
Roedown Research R^2, Davidsonville, MD, USA

Department of Earth Sciences, University College London, Bloomsbury, London, UK

P. L. Guth (ed.), *Military Geoscience*, Advances in Military Geosciences,
https://doi.org/10.1007/978-3-030-32173-4_14

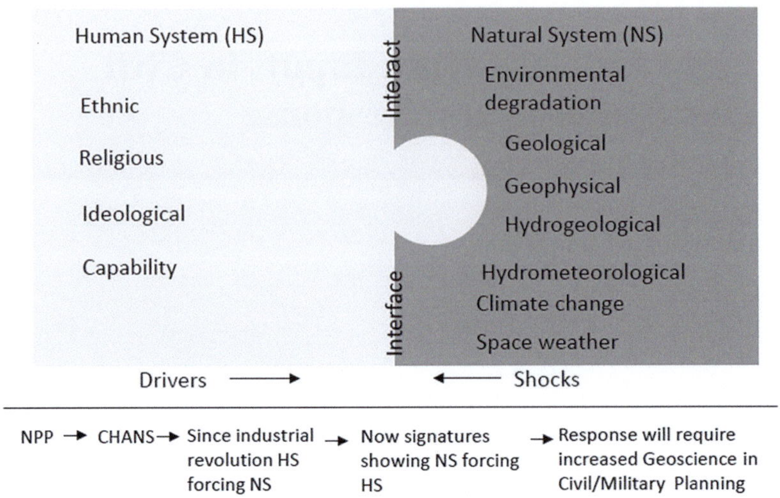

Fig. 14.1 The twenty-first century strategic environment is increasingly best described using the concept of coupled human and natural systems (CHANS) shown here as two connected puzzle pieces. Since the industrial revolution, human systems have been forcing natural systems, but evidence is increasing that natural systems are increasingly shocking human systems by environmental degradation, geological, hydrometeorological, space weather events, and climate change

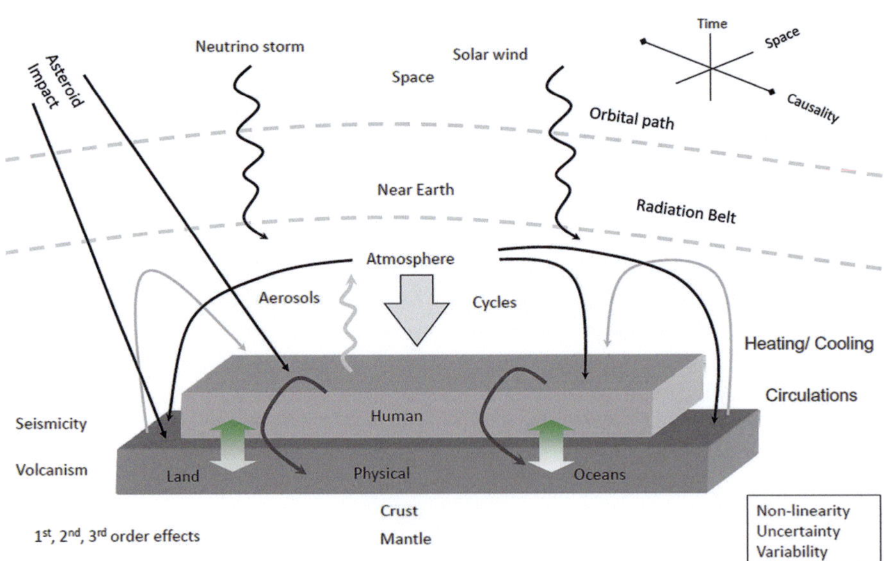

Fig. 14.2 Around the world, increasing evidence that climate change is altering the interfaces and interactions that link human to natural systems is being recognized and here is shown as an evolution from the two-piece puzzle in Fig. 14.1

by understanding factors such as their frequency, duration, and magnitude. Such planning must embrace understanding that natural systems are nonlinear and the past is no longer a guide to the present or the future behavior. Such understanding of the physical environment will need to be obtained using knowledge from geo-eco-bio-physical and technical fields that are collectively described as geoscience inputs. These inputs are especially important when considering the duration of military conflicts and insurgencies and responses to disasters and humanitarian catastrophes. Almost invariably, these last more than one season and are increasingly experiencing the effects of climate change.

Recent policy and direction related to conflict, disaster response, resilience, stabilization (Joint Doctrine Publication n.d.), and humanitarian relief from the UK (UK Government's Humanitarian Policy 2011; UK National Strategy 2010; Committee on Climate Change 2017), the USA (NOAA 2018), UN, NATO [e.g., Joint NATO (n.d.)], World Bank (Bannon and Collier 2003), Intergovernmental Panel on Climate Change (IPCC 2014), and Academia [e.g., (Environmental considerations 2009)] contain recurring themes of environment, natural resources, natural hazards, and climate change. These are identified as drivers of poverty, change, instability, insecurity, cost, and conflict. They occur over variable temporal and spatial scales. The UN Security Council has recognized climate change as a "threat multiplier," exacerbating threats caused by persistent poverty or weak resource management and the possible security implications (UN General Assembly Climate Change 2009). The policy trend is increasing emphasis on prevention. This is demonstrated in efforts in the UN and at national levels to move from a culture of reaction to one of prevention. The emphasis is on sustainable development as a critical contributor to the prevention of conflict, disaster, and human catastrophes. Here we examine how geoscience inputs can assist civil military planning and response. Concepts in selected recent UK, NATO, and UN policies and strategies are highlighted. Areas where geosciences are needed in such policy and direction are explored.

Meeting the Challenge of Climate Change

Since modern record keeping began in 1880, the Earth has experienced sustained higher temperatures (UK Government's Humanitarian Policy 2011; UK National Strategy 2010). Most of the warming occurred in the past 35 years, with 16 of the 17 warmest years on record occurring since 2001. In addition to 2016 being the warmest year on record, 8 of the 12 months (January through September, with the exception of June) were the warmest on record for those respective months. The 2016 rise was the third year in a row to set a new record for global average surface temperatures. 2017 ranked as the second warmest since 1880. The decade 2001–2010 had numerous weather and climate extremes, unique in strength and impact (World Meteorological Organization 2011), and these have continued in the second decade. As greenhouse gas emissions and atmospheric carbon dioxide levels continue to rise, scientists expect the long-term temperature increase to continue (IPCC 2014; Committee on Climate Change 2017; Melillo 2014). Recent warming has been especially strong in Africa, parts of Asia, and parts of the Arctic. The Saharan/Arabian, East African, Central Asian, and Greenland/Arctic Canada subregions have all had 2001–2010 temperatures 1.2–1.4 ° C above the long-term average and 0.7–0.9 ° C warmer than any previous decade. Climate change is increasing the frequency and intensity of natural disasters, particularly hydrometeorological events such as floods, storms, and droughts. It is estimated that 20 years ago, 50% of natural disasters were related to climate, but this figure is now assessed to be nearly 70% (Guha-Sapir and Vos 2009). Globally, the number of reported weather-related natural disasters has more than tripled since the 1960s. In the first decade of the twenty-first century, 2.4 billion people were affected by climate-related disasters compared with 1.7 billion at the end of the last century. The IPCC considers that climate change will likely cause continued increases in the frequency of climate-related hazards, especially floods and storms, with increased incidences of heavier precipitation and stronger winds (IPCC 2014). Together with increasing climate unpredictability, storms and droughts of greater magnitude will expose larger and often less well-prepared regions to the risk of extreme weather events and associated environmental emergencies (e.g., the hurricane season of 2017).

Climate change is causing loss of agricultural land by desertification and flooding of coastal communities and low-lying islands. Sea level rise is increasing the risk to hundreds of millions of people in coastal areas to floods and storms. This is forcing changes in population density and altering access to resources. As a consequence of increased competition for essential resources, migration, political instability, and conflict are likely to rise beyond current levels, especially in at-risk areas which also have growing populations (Global Strategic Trends 2014). Presently, more than half of the world resides in cities, and this will rise to 70% by 2045. The majority of the urban population growth will be concentrated in large and intermediate cities and their informal settlements (Savage and Muggah 2012). Those who lack the capacity to prepare for disasters are the poor, the socially marginalized, women, children, and the elderly (Environmental Emergencies 2009). In adopting the goals of the 2030 Agenda on Sustainable Development and the Paris Agreement on Climate Change, the international community took responsibility for building a sustainable future. But meeting the goals of eradicating hunger and poverty by 2030, while addressing the threat of climate change, will require a profound transformation of food and agriculture systems worldwide (The State of Food and Agriculture 2016). The Tohoku-Pacific Ocean earthquake in 2011 revealed how small increases in seismic and tsunami strengths can overwhelm even developed countries' preparedness capacity both civil and military [e.g., Bulmer (2011)]. The 2017 hurricane season that produced Harvey, Irma, and Maria impacted the Caribbean islands and the US mainland. Comparisons of the resulting impacts demonstrated significant weaknesses in resilience, preparedness, and response occurred in the USA with damage costs estimated to be 265 billion dollars (National Hurricane Center 2018).

Civil Military Planning and Response

At the strategic, operational, and tactical levels and across the full range of military operations, civil military cooperation enables the coordination, synchronization, and de-confliction between military activities and civil actors [e.g., Allied Joint Publication (2013)]. This enables clear linkage between military operations and political objectives. In the mid-1990s, after operations in Bosnia, Herzegovina, and Kosovo, NATO members focused on developing civil military cooperation CIMIC doctrine (Joint Doctrine Publication 2006; Allied Joint Publication-9). The NATO definition of CIMIC is the coordination and cooperation, in support of the mission, between the [NATO] commander and civil actors, including the national population and local authorities, as well as international, national, and nongovernmental organizations and agencies. The UK uses this definition and has integrated it with its *security and stabilization doctrine* (Joint Doctrine Publication 3-40). CIMIC is both a function and a capability and works as a force multiplier. The United Nations uses the term civil military coordination (CMCoord) defined as "the essential dialogue and interaction between civilian and military actors....to protect and promote humanitarian principles, avoid competition, minimize conflict, and when appropriate pursue common goals." US doctrine employs the term civil affairs (CA) defined as "those that enhance the relationship between military forces and civil authorities in areas where military forces are present, to enhance the conduct of civil military operations" (Civil-Military Operations 2013; Department of Defense 2011). The *2010 Joint Operating Environment* lists climate change as one of the security threats the military expected to confront over the next 25 years (The Joint Operating Environment 2010). In 2017 Secretary of Defense James Mattis asserted that climate change is real and a threat to American interests abroad and the Pentagon's assets everywhere, a position at odds with the views of President Trump who appointed him and many in the administration in which he serves.

In the last two decades, many large-scale disasters have occurred in contexts of ongoing conflict or violence, and national and foreign militaries have increasingly been used in disaster response and often in urban settings (e.g., in Haiti after the magnitude 7.0 earthquake in 2010 and in Mexico City after the magnitude 7.1 earthquake in 2017). With an increase in the incidence of natural disasters and conflict, this trend can be expected to continue. Recent experiences of multinational forces responding to disasters and humanitarian catastrophes while conducting warfighting and counterinsurgency and peacekeeping operations have revealed the vast variety of civil and natural contributions and influences that must be considered in today's civil military operations. Such considerations are critical in increasingly multidimensional operations to enable the defined end state to be reached, for the best of the local population, the civil actors, and the military. Within the context of civil military operations, selected recent UK, NATO, and UN policies and strategies are now examined that demonstrate a growing appreciation of the need to recognize and leverage more nonmilitary components of national power. This is achieved using a whole of government, interagency approach including the private sector.

The aim is to derive a holistic and balanced strategy. The focus below is on identifying the need, both recognized and unrecognized, for knowledge on the physical environment and the geoscience implications.

Needs for Geosciences in Recent UK Policy and Strategy

To improve the effectiveness of interventions in complex environments, the UK has pursued efforts to combine civilian and military approaches. The 2006 Comprehensive Approach (Details for a Comprehensive Approach 2006) emphasized the need to promote a shared understanding of the situation, to design structures and processes to respond effectively, and to establish relationships and cultural understanding. This has been a challenge [e.g., The Comprehensive Approach (2010)], but new policy and strategy has been promulgated. This is now examined from the perspective of the UK's need to understand the physical environment as it relates to civil military planning to identify and prevent naturally induced drivers of conflict, disasters, and humanitarian catastrophes.

In 2008, the Cabinet Office released the *National Risk Register* (NRR) (National Risk Register 2008), as part of the British government's *National Security Strategy* with the latest edition released in 2017. It provides an assessment of significant potential risks to the UK divided into natural events, major accidents, and malicious attacks. The first depends upon knowledge of the physical environment for effective geoscience inputs for planning and response. Major accidents and malicious attacks need knowledge of the physical environment depending on their nature, the scale of the affected area, and the duration. The *Adaptation Subcommittee of the Committee on Climate Change* highlighted six thematic risks for the UK (Committee on Climate Change 2017). The top two priorities were flooding and risks associated with coastal changes and risks to health, well-being, and productivity from high temperatures. The NRR provides information on both flooding and heat waves as potential civil emergencies.

The NRR formed the framework for the release of *the National Security Strategy* (NSS) (The National Security Strategy 2010) and the *Strategic Defence and Security Review* (Securing Britain 2010) both released in October 2010. The UK presented its third NSS in November 2015; this time it combined it with its *Strategic Defense and Security Review* to form a single-policy white paper (National security strategy and strategic defence and security review 2015).

The NSS has evolved with each update, but domestic and overseas risks are consistently placed into three tiers. The groups of risks within each tier represent those of highest priority for UK national security looking ahead, taking account of both likelihood and impact. With regard to the requirement for geoscience inputs, tier one lists a growing risk of international military conflict that will draw in the UK. It also lists a natural hazard that requires a national response, while tier two contains an attack on the UK or its overseas territories using chemical, biological, radiological, or nuclear weapons. Tier three contains a large-scale conventional military attack on the UK resulting in fatalities and damage to infrastructure within the UK; a major release of radioactive material from a civil nuclear site within the UK which affects one or more regions; and short- to medium-term disruption to international supplies of resources essential to the UK. It also contains weather and other natural hazards plus environmental events. To a greater or lesser degree, geoscience inputs will be required in planning for all of these and will increasingly need to address manifestations of climate change.

In July 2011, Her Majesty's Government (HMG) published the *Building Stability Overseas Strategy* (BSOS) (Building Stability Overseas Strategy 2011) that takes an integrated medium- to long-term approach promoting stability and prosperity in countries and regions where its interests are at stake. The premise was that addressing instability and conflict overseas was both morally right and in the UK's interest. It linked prevention, early warning, and crisis response. Conceptually there was a long-term notion of stability or "positive peace." The stability that HMG seeks to support through the BSOS can be characterized in terms of representative and legitimate political systems, capable of managing conflict and change peacefully, and societies in which human rights and rule of law are respected, basic needs are met, security established, and opportunities for social and economic development are open to all. This type of "structural stability," which is built on the consent of the population, is resilient and flexible in the face of shocks and can evolve over time as the context changes. This will be achieved by identifying, preventing, and ending instability and conflict overseas, using diplomatic, development, military, and security tools and by drawing on experience, relationships, reputation, and values. There are three mutually supporting pillars: early warning, rapid crisis prevention and response, and investing in upstream prevention. The BSOS makes no mention of the physical environment, but the need to understand it is clear when managing conflict and change peacefully, meeting basic needs, and being resilient and flexible in the face of naturally induced shocks. Consideration has to be given to the geoscience inputs required to confront environmental degradation (Bulmer 2018), conflict over natural resources, natural disasters, and naturally driven humanitarian catastrophes all of which are being influenced by climate change [e.g., Bulmer (2006)].

In September 2011, the revised *UK Governments Humanitarian Policy* (UK Government's Humanitarian Policy 2011; Department for International Development 2006) was released and built on Lord Ashdown's independent *Humanitarian Response Review* (Humanitarian Emergency Response Review 2011) taking account of the BSOS strategy. The UK committed to a multilateral and UN-led and coordinated international humanitarian system with the Department for International Development (DFID) coordinating the UK Government's humanitarian responses. Standard operating procedures have been developed in line with internationally agreed frameworks [e.g., (OCHA (2003, 2007), and Protection of Unarmed Civilians (1999)). Building disaster resilience is a core part of DFID programs helping communities and countries to be better prepared to withstand and rapidly recover from shocks such as an earthquake, drought, flood, or cyclone (Department for International Development 2011). The UK approach to humanitarian assistance has four elements: multilateral, country-specific, directly, and diplomatically. To this end, the UK attaches priority to working with others to support the UN Emergency Relief Coordinator to lead the system and the UN OCHA to fulfill its mandate. At the time that the policy was released, the Secretary of State for International Development recognized that this would require a significant change in the way DFID worked. It is apparent that there is an essential need for knowledge of the physical environment if natural drivers of conflict, disasters, and humanitarian catastrophes are to be understood. Resilience and disaster risk reduction need to be integrated into conflict prevention strategies and climate change adaptation. The DFID policy states that the UK will work with the scientific community and use science, research, conflict analysis, and country knowledge to improve early warning and facilitate early action. No specifics are given, but it is stated that the Chief Scientific Advisor's network will be drawn upon to improve the use of science in both predicting and preparing for disasters. The desire to predict natural disasters is understandable, but at present science does not support it consistently. With regard to the easier issue of preparing for potential natural- and human-induced disasters, the policy recognizes that building disaster resilience requires improvements to social, economic, environmental, political, and physical planning.

BSOS is still HMG government policy and was integrated into the *2015 Strategic Defense and Security Review*, but in 2017, the *Building Stability Framework* (BSF) outlined five building blocks and five shifts for DFID (Building Stability Framework 2016). These relate to its work on building stability in fragile states and societies. The five building blocks are fairer power structures; more inclusive economic development; better mechanisms for resolving conflict; more effective and legitimate institutions; and a more supportive regional environment. Compared to the BSOS, the BSF is less prescriptive. Again it makes no mention of the physical environment, but it is clear that the framework covers a spectrum of fragility and that understanding of the physical environment is critical to its success. The BSOS and BSF need further examination as to where the geoscience expertise resides; what capabilities exist in the UK especially in prediction, hierarchies of expertise, and expert decision-makers; and where the interfaces exist in civil military planning.

Needs for Geosciences in Recent NATO Policy and Strategy

*NATO's Strategic Concep*t (Active Engagement, Modern Defence, 2010), released in November 2010, recognizes a much wider range of threats to international security than existed hitherto. In addition to continuing to provide for collective defense, the concept states that the Alliance must stand ready "to contribute to effective conflict prevention and to engage actively in crisis management, including crisis response operations." NATO CIMIC is applicable to both Article 5 Collective Defense and Non-Article 5 Crisis Response operations and has increasingly been used in this latter role (NATO's Role in Disaster Assistance 2001). The Strategic Concept goes on to state "The interaction between Alliance forces and the civil environment (both governmental and non-governmental) in which they operate is crucial to the success of operations" (Allied Joint Publication-9). It commits the 28-member alliance to use its political and military capabilities to prevent crises, manage conflicts, and stabilize post-conflict situations and to a broad spectrum of activities. These includes enhancing integrated civilian-military planning throughout the crisis spectrum; forming an appropriate but modest civilian crisis management capability to interface more effectively with civilian partners; and developing the capability to train and develop local forces in crisis zones. It is clear that to contribute to conflict prevention and engage in crisis management and response, NATO civil military planners require understanding of the physical environment and climate change. Expertise and capability continues to be established inside NATO and needs to continually be examined to determine how it is best interfaced into the joint force commander's civil military staffs that are fully integrated into the headquarters.

Needs for Geoscience in Recent UN Policy and Strategy

Recent UN policy and guidelines require civil military planners to address undesired and often unknown environmental legacies of UN missions and to make them role models regarding environmental stewardship (Environmental Guidebook for Military Operations 2008; UNDG 2013; United Nations 2008; DPKO 2009; Waleij and Liljedahl 2009; Greening the Blue Helmets 2012). The *Rio Declaration on Environment and Development* (Rio Declaration on Environment and Development 1992) produced at the 1992 United Nations Conference on Environment and Development called for States to "respect international law providing protection for the environment in times of armed conflict and cooperate in its further development, as necessary." The UN Environment Programme recognizes that in the last 60 years, at least 40% of all intrastate conflicts have a link to natural resources (Greening the Blue Helmets 2012; Halle 2009). High-value resources include timber, gold, minerals, and oil as well as scarce ones like fertile land and water. This link doubles the risk of a conflict relapse in the first 5 years of peace. Since 1990, at least 18 violent conflicts have been fuelled by exploitation of natural resources (Bannon and Collier 2003). An examination of conflict drivers reveals that competition over natural resources can contribute to the outbreak of conflict, financing, and sustaining of conflict. Control over revenue generating natural resources can also undermine peacemaking efforts. From this perspective conflict and insurgency can be seen to be an economic activity not just political or ideological (Bulmer 2018; Bulmer 2019). Since 1948, 20% of UN peacekeeping missions have had a direct or indirect mandate to address natural resources (Fig. 14.3) but only a few to help the host country better manage its natural resources. Peacekeeping operations have important implications on natural resource and the potential for significant impacts on the environment. Natural resources are often a fundamental aspect of conflict resolution, livelihoods, and confidence-building at the local level. Indeed addressing the risks and opportunities presented by natural resources is often critical to the success of

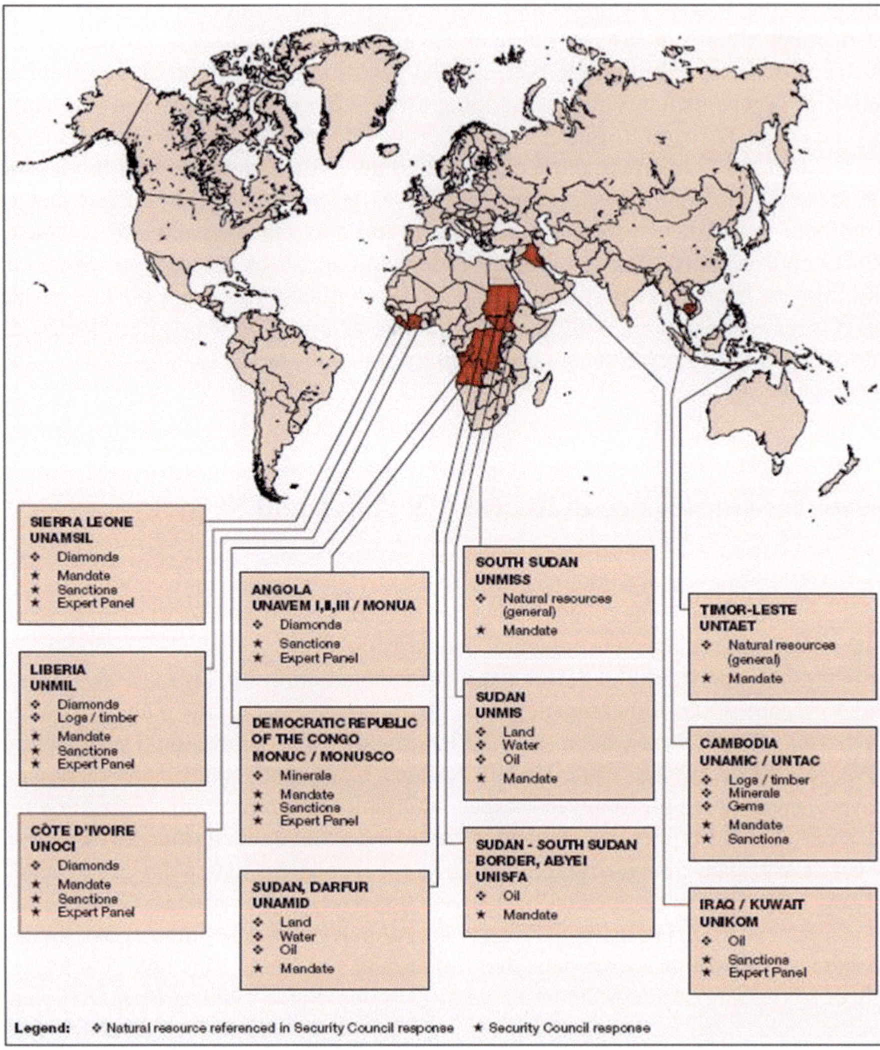

Fig. 14.3 Security Council responses to conflicts linked to natural resources from 1948 to 2011. From Greening the Blue Helmets, UNEP, 2012

UN peacekeeping efforts (Halle 2009). This also relates to the provision of "a safe and secure environment" mandated under *Protection of Civilians* and can no longer be seen as distinct from the maintenance of peace and security (Protection of Unarmed Civilians 1999).

UN peacekeeping missions that have a direct or indirect mandate to address natural resources require in-depth understanding of the physical environment to effectively plan and analyze risks and opportunities to derive sustainable solutions. At the field level, the Security Council typically establishes a Panel of Experts (also

Table 14.1 Conflicts in the last 60 years with a link to natural resources

Country	Duration	Resources
Afghanistan	1978–2001 (23 years)	Gems, timber, opium
Angola	1975–2002 (27 years)	Oil, diamonds
Burma	1949–present	Timber, tin, gems, opium
Cambodia	1978–1997 (19 years)	Timber, gems
Colombia	1984–present	Oil, gold, coca, timber, emeralds
Congo, Dem Rep. of	1996–1998, 1998–2003, 2003–2008	Copper, coltan, diamonds, gold, cobalt, timber, tin
Congo, Rep. of	1997–present	Oil
Côte d'Ivoire	2002–2007 (5 years)	Diamonds, cocoa, cotton
Indonesia–Aceh	1975–2006 (31 years)	Timber, natural gas
Indonesia–West Papua	1969–present	Copper, gold, timber
Liberia	1989–2003 (14 years)	Timber, diamonds, iron, palm oil, cocoa, coffee, rubber, gold
Nepal	1996–2007 (11 years)	Yarsa gumba (fungus)
PNG–Bougainville	1989–1998 (9 years)	Copper, gold
Peru	1980–1995 (15 years)	Coca
Senegal–Casamance	1982–present	Timber, cashew nuts
Sierra Leone	1991–2000 (9 years)	Diamonds, cocoa, coffee
Somalia	1991–present	Fish, charcoal
Sudan	1983–2005 (22 years)	Oil

Adapted from Bannon and Collier 2003 (Joint NATO n.d.)

known as Groups of Experts or Expert Panels) (Boucher and Holt 2009). These are small, civilian, fact-finding teams that advise on the scope, monitor the effectiveness, and report on the implementation of any sanctions on countries, individuals, or groups who threaten peace and security (Greening the Blue Helmets 2012). The Panels can also investigate violations of UN sanctions, as well as offer analysis on the nature of the conflicts, the exploitation of natural resources, and the grounds for lifting sanctions. Addressing land and natural resource challenges is also becoming more common within the activities of Civil Affairs, which are civilian components of UN peacekeeping operations that work at the social, administrative, and subnational political levels to facilitate the countrywide implementation of peacekeeping mandates (Table 14.1).

Duration of Involvement

The relevance of geoscience inputs to civil military planning and cooperation is directly related to the type of mission or operation, its complexity, and its geography. Missions must now be geospatially and temporally contextualized within an understanding of CHANS. The duration required to achieve planned strategic, operational or tactical end states for military conflicts, and responses to a disaster or humanitarian catastrophes must increasingly be informed by a deeper understanding of the physical environment and climate change. This must be combined with the rising incidence of conflicts (Trends in Armed Conflict 2016), natural disasters and resulting costs (Global Catastrophe Recap 2018), and increasing competition for natural resources. Table 14.2 provides some historical perspective on the durations of a range of military operations over the period 1948 to 2018. The involvement of Security Council responses to conflicts related to natural resources ranges from 2 years (East Timor) to 19 years (DRC) with the largest number lasting between 5 and 10 years and the mean being 9.1 years. An examination of 208 UN peacekeeping operations shows that the durations range of 1 to 11 years (MONUC in the DRC) with a mean of 3.7 years. The mean duration of operations in Africa,

Table 14.2 Durations of a range of military operations

Description	0–5 years	5–10 years	10–15 years	15–20 years	25–30 years	Total military operations	Mean years
UN conflict linked to natural resources	1	6	4	1		12	9.1
UN peacekeeping operations in Africa	19	3	1			23	3.3
UN peacekeeping operations in Americas	9		1			10	3.1
UN peacekeeping operations in Asia	8	2				10	2.4
UN peacekeeping operations in Europe	4	3		1		8	5.5
UN peacekeeping operations in the Middle East	4	1	2			7	5.1
Wars involving NATO (1945–present)	12	3	1	1		17	3.9
Wars involving UK (1939–present)	25	5	3		1	34	4.9
Insurgencies involving UK (1948–present)	8	3	3		1	15	8.0

Durations were calculated from data available on http://en.wikipedia.org/wiki/List_of_United_Nations_peacekeeping_missions accessed 26 March 2018. https://en.wikipedia.org/wiki/List_of_wars_involving_the_United_Kingdom; accessed 26 March 2018. https://en.wikipedia.org/wiki/List_of_NATO_Operations accessed 26 March 2018 the Americas, Asia, Europe, and the Middle East is 3.7 years, but each continent has at least one mission that has continued over 5 years. In the case of Africa, it is MONUC in the DRC (11 years), UNMEE in Ethiopia and Eritrea (8 years), UNAMSIL in Sierra Leone (6 years), and UNMIS in Sudan (6 years); for the Americas it is MINUSTAH in Haiti (13 years); for Asia it is UNMIT in Timor-Leste (6 years) and UNMOT in Tajikistan (6 years); for Europe it is UNOMIG in Georgia (16 years), UNMIBH in Bosnia and Herzegovina (7 years), UNPREDEP in Macedonia (7 years), and UNMOP in Croatia (6 years). Wars involving NATO (1945 to present) range from 1 month for the 1995 air campaign over Bosnia to 19 years for KFOR in Kosovo with a mean of 3.9 years. The International Security Assistance Force was in Afghanistan for 13 years. A total of 34 wars involving the UK (1939 to present) were examined and range from 1 year (2011 Libyan Civil War) to 30 years (The Troubles, Northern Ireland) that have a mean of 4.9 years. Insurgencies involving the UK (1945 to present) range from 1 year (e.g., Eritrea 1949) to 30 years (The Troubles, Northern Ireland) that have a mean of 8 years. This mean is high due to insurgencies in Palestine, Dhofar in Oman, and Afghanistan all lasting 13 years, Malaya (12 years), Egypt (10 years), and Kenya (8 years). In all instances the mean durations of operations are multi-year. These previous experiences underscore the need for future operational plans to incorporate understanding of the changing physical environment in the area of operations and likely future issues. These include environmental degradation, land use change, geological and hydrometeorological hazards, space weather, climate change, natural disaster, and environmental emergencies. For ongoing civil military operations and those in the future, it is paramount that additional consideration is given to matching mission timelines in operational plans with knowledge of the likely changes in the physical environment and how climate change will affect them over the identified period. This can capitalize on the continually evolving geoscience modeling and improvements in the type and resolution of remotely sensed data.

Discussion

Examination of the policies and strategies above reveals the need for improved understanding of the physical environment if they are to succeed. Although the aims, objectives, intents, and aspirations are laid out in the policies and strategies, they lack details for how to access geoscience knowledge, capability, and training related to the physical environment. Within military planning structures, there is often no geosciences expertise beyond geospatial mapping and engineering. This has made it very difficult to integrate contingency planning for resource management, environmental disaster, or climate change into missions or operations.

There is a view across NATO that military headquarters are too large and too complex, operating at tempos too high for regular staff to become deeply versed in areas of expertise. One way to overcome this is to access expertise from outside, and Fig. 14.4 shows a conceptual model to enable understanding of the environmental

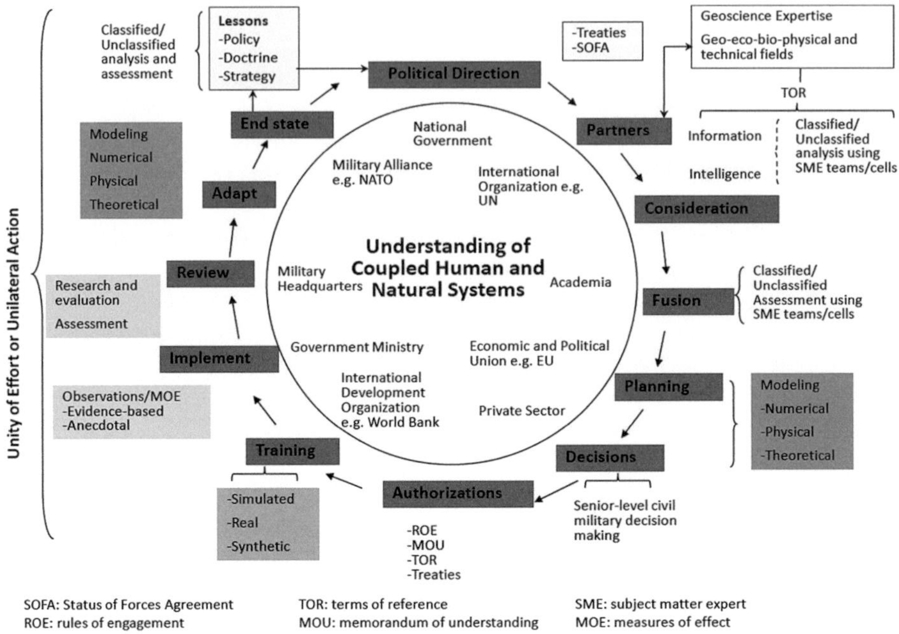

Fig. 14.4 A model for accessing knowledge, capability, and training in areas regarding the physical environment to understand the strategic, operational, and tactical significance of naturally induced drivers of conflict, disasters, and humanitarian catastrophes in coupled human and natural systems. The model operates as a cycle turning clockwise

context in CHANS, whether strategic, operational, or tactical. This promotes action in the face of early warning of change and delivers critical insight into drivers of conflict, paths to stability, reconstruction, and sustainable peace. The model operates as a cycle, and clockwise motion will be initiated by the political direction that will define whether the mission or operation is multi- or unilateral and the level of civil military cooperation. This will be informed by existing treatises and status of forces agreements (SOFAs). Once the political direction is given, pathways to geosciences partners must be established. Such partners may be individuals as well as teams of subject matter experts (SMEs) in the geo-eco-bio-physical and technical fields. Critical will be access to geoscientists working in the information as well as the intelligence arenas. This will require early establishment of terms of reference (TORs) and protocols for working as part of the cycle. As these partners are organized into working groups, consideration of CHANS must focus not just on the physical process or mechanics but on cause and effect of natural events (e.g., floods, earthquakes etc.) and first-, second, and third-order consequences, how they differ, diverge, and link over a range of time, scales, and at variable rates. This approach will enable analysis to be integrated into infrastructure assessments (Bulmer 2015).

Geoscience partners need to connect to analysts trained in the formal methods and modes used within organizations to convey information. This enables information collation in a form that can be passed to planning staff. Workable memorandums of understanding (MOUs) that enable sharing of classified and unclassified intelligence and technical geoscience inputs and geospatial data are critical to achieving the fusion of considerations derived from different analysis and assessment teams. Fusion of products and analysis from all partners can be achieved using coordinated teams or cells of SMEs who provide consensus assessments to civil military planners. These enable planners to produce a dynamic common operating picture (COP) that can be analyzed in a geographic information system architecture (GIS) resulting in realistic and well-informed options matched with doctrine [e.g., Joint Doctrine Publication 3-70 Battlespace Management (2008)], policy, and strategy aims. Scenarios derived from these options can then be modeled using theoretical, numerical, and physical approaches.

The level at which actions in any military operation are undertaken will need to synchronize in time and space to achieve greatest effect. In CHANS interactions, interfaces, drivers, and effects occur through, across, and within these dimensions. This highlights the challenges confronting decision-makers when nonlinearity, uncertainty, and variability associated with natural systems are added into CHANS planning. For military operations this increases the complexity but critically will expand awareness and understanding. This in turn will increase the range of options and effects that can be conceived and modeled.

The primary responsibility of decision-makers is to ascertain variables and interactions in CHANS allowing for manipulation and amelioration if not solutions. The paradigm here is "what needs to be done" rather than what the civil military organizations "can do." Using the COP civil and military leaders can make geoscience-informed decision, and senior-level decision-makers can provide authorizations with rules of engagement (ROEs) that deal with the CHANS interactions and are set within SOFAs, MOUs, and treaties.

Following authorizations, the cycle next connects with civil military actors selected to train for the mission/operation and implement the plan. All levels of actors from top to bottom need to become familiar with the concepts of CHANS and environmental guidelines. Those responsible for addressing natural resources, environmental crimes and degradation, natural disasters, and natural catastrophes require significant understanding of the physical environment to effectively plan, analyze, and address natural resource risks and opportunities. They need to be trained in evidence-based and anecdotal data collection, change detection, analysis, and assessment methodologies used in geoscience and technical fields. Use can be made of templates such as those contained in policy [e.g., Environmental Guidebook for Military Operations (2008), and Guidelines for Environmental Emergencies (2005)] and in the NATO CIMIC reporting system. Wherever possible, training should include how to build capacity in the host nation for monitoring and enforcement of environmental regulations. Such training can utilize the geospatial data collated in the fusion cell capitalizing on its dynamic attributes.

Providing the appropriate level of geoscience-related training is a complex undertaking and must be matched with the mission/operation timelines, budgets, mandates, manpower, and capabilities. A range of training strategies in real, simulated, and synthetic environments should be considered. Significant savings and benefits can be obtained using unity of effort (e.g., UN Cluster IASC (2006) and military headquarters), but there are innumerable reasons why countries, militaries, government departments, international organizations, and nongovernmental organizations may conduct themselves unilaterally.

During implementation of civil military operational plans, observations of CHANS and measures of effectiveness (MOEs) in implementation must cover natural resources, environmental crimes and degradation, natural disasters, and natural catastrophes. Whenever possible MOEs should be designed in agreement with all civil military actors and covered under MOUs, TORs, and SOFAs. MOE's must enable evidence-based reviews of progress (e.g., ISAF Afghan Country Stability Picture 2012) that holds information on different Afghan National Development Strategy sectors such as education, good governance, health agriculture and rural development, infrastructure, and natural resources). This enables research and evaluation to be undertaken [e.g., DFID (2012), and Bulmer (2012)] that results in an evolving understanding of the changing interface and interactions that link CHANS. A review cycle of assessments from ongoing missions/operations enables plans to be adapted. As in the initial planning phase of the CHANS cycle, adaptations can be modeled using numerical, physical, and theoretical techniques. When assessments and MOEs reveal achievement of strategic, operational, and tactical CHANS-related end states, then classified and unclassified lessons can be derived that can be related to the physical environment. These inform policies, doctrine, and strategies for future CHANS inputs into missions and operations.

Conclusions

The twenty-first-century world is increasingly best described by coupled human and natural systems. Within the natural system, environmental degradation, geological and hydrometeorological hazards, space weather, and climate change all interface and interact with ethnic, religious, ideological, and capability drivers creating the human systems. Around the world evidence is increasing that climate change is altering the interfaces and interactions that link human to natural subsystems. This is being recognized in civil and military policies and strategies. With the incidence of conflict and natural disasters increasing, national and foreign militaries can be expected to play a bigger role working alongside civil actors. Increasingly this will be in degraded environments and in urban settings. This requires a renewed emphasis on identifying and preventing naturally induced drivers of conflict, environmental crimes, disasters, and humanitarian catastrophes. For civil military mission/operations, this can be achieved using a conceptual model that works in a cycle to obtain understanding of the environmental context in CHANS, whether strategic, operational, or tactical.

This cycle provides a framework for accessing geosciences knowledge, capability, and training in areas regarding the physical environment and CHANS. The relevance of geoscience inputs to the strategic, operational, and tactical level civil military planning and cooperation is directly related to the type of operation, its complexity, and its geography. Going forward this must now be contextualized within the understanding of CHANS that exists in the defined time and space of the mission/operation. Geoscience inputs have critical strategic, operational, and tactical value when considering the duration of past military conflicts, responses to disasters, and humanitarian catastrophes.

In the future, civil military planners and responders must have a greater depth of understanding of the changing physical environment and climate change when conceiving, planning, and implementing sustainable solutions in societies weakened by conflicts, disasters, or humanitarian catastrophes. Concepts of nonlinearity, uncertainty, and variability in natural systems must be embraced. Governments and militaries must identify where advice and geoscience expertise resides when planning early warning and early action. They must use modeling to examine the long timeframes required to meet threats upstream, undertake stabilization, and achieve stability in conflicts, disasters, environmental emergencies, and humanitarian catastrophes. This demands a continuous and uncompromising assessment mechanism to measure the ability of militaries and civil actors to deliver across these areas of civil military operation.

References

A secure and prosperous United Kingdom. National security strategy and strategic defence and security review. (2015). H M Government, London. p. 96.

A strong Britain in an age of uncertainty: The National Security Strategy. (2010). H M Government, London. p. 39.

Active Engagement, Modern Defence. NATO's Strategic Concept. (2010). NATO Public Diplomacy Division, Brussels, Belgium. p. 40.

Afghan Country Stability Picture. NATO International Security Assistance Force. (2012). https://www.slideshare.net/BZjoe/20120917-u-acsppme

Allied Joint Publication 3.4.9. Allied Joint Doctrine for Civil-Military Cooperation Edition A, version 1. NATO Standardisation Agency AJP-3.4.9. (2013). See also Joint Publication 3-57 (2008). Civil Military Operations. Joint Chiefs of Staff, Washington, DC. p. 185.

Allied Joint Publication-9 NATO Civil-Military Co-operation (CIMIC) Doctrine. p. 57. NATO CIMIC Policy MC 411/1 (n.d.).

Bannon, I., & Collier, P. (Eds.). (2003). *Natural resources and violent conflict* (p. 429). Washington, D.C.: World Bank. ISBN 0-8213-5503-1.

Boucher, A., & Holt, V. K. (2009). *Targeting spoilers: The role of United Nations panel of experts. Stimson center report no. 64* (p. 206). Washington, D.C.: Stimson Center.

Building Stability Framework. (2016). Department for International Development. London. p. 28. http://www.international-alert.org/sites/default/files/DFID_Building%20stability%20framework%202016.pdf

Building Stability Overseas Strategy. (2011). Department for International Development, Foreign and Commonwealth Office, Ministry of Defence, London. p. 40.

Bulmer, M. H. (2006) Knowledge of physical process can assist natural disaster response. *Monday Developments, 24*, 6, 10–11.

Bulmer, M. H. (2011). Worst-case scenario: The Tohoku-Pacific Ocean Earthquake. Royal United Services Institute, Analysis, Commentary, ref:C4D99EFC6CE5B5. http://www.rusi.org/analysis/commentary/ref:C4D99EFC6CE5B5/

Bulmer, M. H. (2012). *Haitian earthquake: Civil military cooperation two years on. RUSI analysis*. London: The Royal United Services Institute.

Bulmer, M. H. (2015). *Geological and infrastructure assessment report (GIAR). A study of the metropolitan manila earthquake contingency plan: In support of 77 brigade exercise civil bridge 15A* (p. 53). Nottingham: Chilwell.

Bulmer, M. H. (2018). Military use of environmental degradation by Islamic State, Southern Nineveh, Iraq. Scientia Militaria, South African Journal of Military Studies, Vol 46, Nr 1, 2018. https://doi.org/10.5787/46-1-1228.

Bulmer, M. H. (2019). Environmental Degradation in the Battlespace. British Army Review, BAR Special Report, Culture in Conflict. The Learning from Conflict Series, 48–60.

Civil-Military Operations. Joint Publication 3-57. (2013). p. 173. https://www.hsdl.org/?view&did=748621

Climate change impacts in the United States: The third national climate assessment (2014) In J. M. Melillo, T. C. Richmond, & G. W. Yohe (Eds.), U.S. Global Change Research Program, p. 841. https://doi.org/10.7930/J0Z31WJ2.

Defining disaster resilience: A DFID approach paper. (2011). Department for International Development, London. p. 20.

Department of Defense Support to Foreign Disaster Relief. Hand book for JTF Commanders and Below. (2011). GTA 90-01-030. p. 314.

Details for a comprehensive approach. (2006). Stabilization Unit, London. p. 2.

Environmental considerations for building peace. (2009). *The Pearson Papers*, *12*, p. 117.

Environmental Emergencies. (2009). *Learning from multilateral response to disasters* (p. 89). Geneva: Joint UNEP/OCHA Environmental Unit.

Environmental Guidebook for Military Operations. (2008). Developed by multinational working group consisting of representatives from the defense organizations of Finland, Sweden, and the United States. p. 118.

Environmental Policy for UN Field Missions, DPKO. (2009). New York: The Department of Peacekeeping Operations and the Department of Field Support. https://peacekeeping.un.org/en/environmental-impact-and-sustainability

From conflict to peacebuilding. The role of natural resources and the environment. (2009). In J. Halle (Ed.), United Nations Environment Programme, Kenya, p. 50. ISBN: 978-92-807-2957-3.

Global Catastrophe Recap. (2018). AON Benfield. Impact Forecasting. January, p. 14.

Global Strategic Trends – Out to 2045. Fifth Edition. (2014). Strategic Trends Programme, Ministry of Defense, Development, Concepts and Doctrine Centre, Shrivenham, Swindon. p. 202.

Greening the Blue Helmets. (2012). *Environment, natural resources and UN peacekeeping operations* (p. 134). Nairobi, Kenya: United Nations Environment Programme.

Guha-Sapir, D., & Vos, F. (2009). Quantifying global and environmental change impacts: Methods, criteria and definitions for data on hydro-meteorological hazards. In Bausch, H. G., et al., (Eds.), *Coping with global environmental change, disasters and security threats* (p. 1815). Berlin: Springer-Verlag. ISBN 978-3-642-17776-7.

Guidelines for Environmental Emergencies. (2005). *Joint UNEP/OCHA Environment Unit* (p. 34). Geneva: United Nations.

Guidelines on the use of military and civil defense assets to support UN humanitarian activities in complex emergencies. (2003). Office for the Coordination of Humanitarian Affair (OCHA). p. 21.

Humanitarian Emergency Response Review. (2011). Chaired by Lord Ashdown. p. 76.

IASC, Guidance note on using the cluster approach to strengthen humanitarian response. (2006). Inter-agency Standing Committee, p. 15.

IPCC: Climate change 2014: Synthesis report. (2014). Contribution of Working Groups I, II and III to the Fifth Assessment Report of the Intergovernmental Panel on Climate Change [Core Writing Team, Pachauri, R. K. & Meyer, L. A. (Eds.)]. IPCC, Geneva, Switzerland, p. 151.

Joint Doctrine Publication 3-90 Civil Military Co-operation (CIMIC). (2006). Development, Concepts and Doctrine Centre. Shrivenham, Swindon. p. 46.

Joint Doctrine Publication 3-40 Security and Stabilization. Development, Concepts and Doctrine Centre. p. 265. It builds on the US Field Manual (FM) 3-24 Counterinsurgency and addresses the challenge of setting the conditions for enduring stability (n.d.).

Joint Doctrine Publication 3-70 Battlespace Management. (2008). *Development* (p. 51). Shrivenham, Swindon: Concepts and Doctrine Centre.

Joint NATO Doctrine for Environmental Protection during NATO Led Military Activities, NATO, STANAG 7141 EP. See also Joint NATO Waste Management Requirements During NATO Led Military Activities, NATO, STANAG 2510 (n.d.).

Marina, A., Asbjornsen, H., Baker, L. A., Brozovic, N., Drinkwater, L. E., Drzyzga, S. A., et al. (2011). Research on coupled human and natural systems (CHANS): Approach, challenges, and strategies. *Bulletin of the Ecological Society of America, 92*, 218–228. https://doi.org/10.1890/0012-9623-92.2.218.

National Hurricane Center. Costliest U.S. tropical cyclones table updated. (2018). National Oceanic and Atmospheric Administration. p. 3. https://www.nhc.noaa.gov/news/UpdatedCostliest.pdf

National Risk Register. (2008). Cabinet Office, London. p. 39.

NATO's Role in Disaster Assistance. (2001). NATO Civil Emergency Planning Euro-Atlantic Disaster Response Coordination Centre, Brussels, Belgium. p. 57.

NOAA National Centers for Environmental Information, State of the Climate: Global Climate Report for Annual 2017, published online January 2018, retrieved on March 17, 2018 from https://www.ncdc.noaa.gov/sotc/global/201713

Oslo guidelines on the use of foreign military and civil defense assets in disaster relief (2007). UN Office for the Coordination of Humanitarian Affairs (OCHA). p. 32.

Our common journey: A transition toward sustainability. (1999). National Research Council Policy Division Board on Sustainable Development. Washington, DC: Natl. Acad. Press, p. 384: https://doi.org/10.17226/9690.

Promoting innovation and evidence-based approaches to building resilience and responding to humanitarian crises, DFID (2012). London: Department for International Development, p. 44.

Protection of Unarmed Civilians. United Nations Security Council. S/PRST/1999/6. 12 February (1999). http://www.un.org/en/ga/search/view_doc.asp?symbol=S/PRST/1999/6

Rio Declaration on Environment and Development. (1992). United Nations Conference on Environment and Development (UNCED), Rio de Janeiro. A/CONF.151/26 (Vol. I). http://www.un.org/documents/ga/conf151/aconf15126-1annex1.htm

Savage, K., & Muggah, R. (2012). Urban violence and humanitarian action: Engaging the Fragile City. *The Journal of Humanitarian Assistance*. http://sites.tufts.edu/jha/archives/tag/urban-violence

Saving lives, preventing suffering and building resilience. The UK Government's Humanitarian Policy. (2011). London: Department for International Development, p. 24.

Saving lives, relieving suffering, protecting dignity: DFID's Humanitarian Policy. (2006). Department for International Development, London. p. 30.

Scale and geographic inquiry: Nature, society, and method. In E. Sheppard, & R. B. McMaster (Eds.), (2004). Wiley-Blackwell. p. 288. ISBN 0-631-23070-X.

Securing Britain in an age of uncertainty: The strategic defense and security review. (2010). H M Government, London. p. 75.

Stern, N. (2007). *The economics of climate change: The stern review* (p. 576). Cambridge: Cambridge University Press.

The comprehensive approach: The point of war is not just to win but to make a better peace. (2010). Seventh Report of Session 2009–10. House of Commons Defence Committee. The Stationary Office Limited, London. p. 249.

The Joint Operating Environment. (2010). United States Joint Forces Command. p. 76. https://fas.org/man/eprint/joe2010.pdf

The State of Food and Agriculture. (2016). *Climate change, agriculture and food security* (p. 194). Rome: Food and Agriculture Organization of the United Nations. http://www.fao.org/3/a-i6030e.pdf

Trends in Armed Conflict, 1946–2015. Conflict Trends 08 (2016). Oslo: Peace Research Institute. p. 4. www.prio.org/ConflictTrends

UK Climate change risk assessment synthesis report: Priorities for the next five years. (2017). Committee on Climate Change, July 2016. Holbein Place, London. p. 86.

UK National Strategy for the Protection of Civilians in Armed Conflict. (2010). Foreign and Commonwealth Office, UK Aid, and Ministry of Defense. p. 16.

UN General Assembly Climate Change and its Possible Security Implications. (2009). Report of the Secretary General, Geneva. United Nations Security Council Resolution. UN 09-50946. p. 29.

UNDG Guidance note on natural resources and land in transition settings. (2013). United Nations Development Group. p. 38.

United Nations peacekeeping operations: Principles and guidelines (2008). UN Department of Peacekeeping Operations and UN Department of Field Support, New York. https://peacekeeping.un.org/en/principles-of-peacekeeping

Waleij, A., & Liljedahl, B. (2009). The military as environmental steward in peace operations. Pearson papers volume 12. FOI-S--3282--SE www.pearsoncentre.org/dl69

Weather extremes in a changing climate: Hindsight on foresight (2011). World Meteorological Organization, 1075. ISBN: 978-92-63-11075-6. p. 20.

Chapter 15
Insurgency and the City

Andrew D. Lohman

Introduction

In many ways the current era resembles the 1950s and 1960s, which Sanger (1970) termed "the Insurgent Era." Wars between states remain relatively rare events, while struggles to overthrow governments or ruling authorities continue to be the most prevalent form of warfare today (Harbom and Wallenstein 2007, Black 2016) and "probably been the most prevalent type of armed conflict since the creation of organized political communities" (O'Neill 2005, p.1). Media coverage of the conflicts in Iraq, Afghanistan, Syria, Libya, Mali, and the DRC and numerous other recent and ongoing struggles lend evidence to these claims. And similar to the period of which Sanger wrote, these conflicts have spurred a tremendous body of scholarly and popular literature to uncover patterns and processes to better understand how these conflicts take shape and ways to preclude or counter them.

Perhaps one of the more challenging aspects of research in this area stems from the lack of consensus on definitions, as varying scholars have proposed a myriad of criteria by which to define and categorize conflicts within states and struggles to wrest power from incumbent regimes (O'Neill 2005; Black 2016). For purposes of this analysis, an insurgency is defined as "a violent political process through which peoples or groups, who are excluded from power, contest the ruling authority to alter or replace existing power relationships" (Lohman and Flint 2010, p. 1154). This chapter, however, does not seek to provide an in-depth discussion of the nuances of definitions, but rather focuses on the geographic or spatial aspects of insurgencies and the role that cities and urban settlements play in the course of such conflicts.

Almost all inquiry into insurgencies continues to be the work of historians and political scientists, yet virtually all of these works discuss, to varying degrees, the

A. D. Lohman (✉)
Department of Geography and Environmental Engineering, West Point, NY, USA
e-mail: andrew.lohman@usma.edu

P. L. Guth (ed.), *Military Geoscience*, Advances in Military Geosciences,
https://doi.org/10.1007/978-3-030-32173-4_15

role and impact of geography on the course and outcome of these conflicts, including cities. Yet like "insurgency," most people have an idea of what the terms "city," "urban," and "rural" mean or imply, but within the literature on insurgencies, these are vaguely defined, if at all (Hills 2004). Likewise, this chapter does not seek to address this shortcoming, but to provide a survey of the literature on insurgencies and cities. Although vaguely defined, cities – and urban settlements of all sizes – have likewise received varying degrees of scholarly attention. Yet today, with over half the world's population living in urban areas, steadily increasing urbanization rates, and growth of urban settings (in spatial size and population), suggest that earlier calls announcing the 'urbanization of insurgency' are coming true (O'Sullivan 1991; Taw and Hoffman 1994; Kilcullen 2013).

Cities in Insurgencies

Throughout history, cities have featured prominently in insurgency struggles, and early Marxist perspectives on revolution viewed cities as the base of industry, the primary sites in which workers would unite and topple the capitalist political structure (Oppenheimer 1969). The Cold War period saw an explosion of research and publications on insurgency, and guerilla warfare as colonies around the world sought independence or the overthrow of incumbent regimes through violence, a large proportion of which were driven by Marxist ideology. Although much of this research focused largely on guerrilla warfare – the dominate form or method of waging an insurgency at the time, defined here as hit and run tactics by small, semi-organized groups that conduct harassment-type attacks and then usually disperse or retreat following a military operation (Laqueur 1976; Asprey 1994; O'Neill 2005) – much of what we know and conceive about insurgency as a process emerged during this period.

Initial efforts by Mao in China and Ho Chi Minh in Vietnam to seize power and control over the state were, following the Marxist prescription, also focused primarily within the major cities. However, Chinese nationalist and French colonial forces were able to thwart these efforts, and the revolutionary movements were forced to continue their struggles in rural areas. Indeed, the Chinese Communist Party's success and Mao's theories on using guerrilla warfare to achieve political objectives became the model and basis for numerous subsequent insurgencies around the globe (Marks 2003). Cities, in these examples, proved difficult, if not impossible, operating environments for nascent insurgent movements.

Mao's success and his theories on mobilizing the real strength of China at the time – the rural peasantry – ushered in a "new" approach to insurgency. Often termed the rural approach, Mao's prescriptions sought to mobilize the rural populations of China (estimated at 60–70% of the population at the time), establish and expand rural bases areas, and "encircle the cities" (Mao 1961, 1967). In this way, the insurgency was largely and predominantly focused on securing the countryside with the cities, and ultimately the capital, as the final objectives which would ultimately topple the incumbent regime and secure control over the state. Although history is replete with examples of rural-based movements, Mao's contributions to theoretical understandings of insurgency as a process, which proceeds through distinct and subsequent phases or stages, have had a profound effect on our understandings of insurgency as a political process (Sanger 1970; O'Neill 2005). Mao's prolific writings further highlighted the salience of geographic and spatial factors in the course of such movements.

Within the vast literature that emerged during this period, geographic factors have, to varying degrees, been noted to play an often crucial role in the course of such conflicts. Relatively few geographers, though, undertook research on guerilla warfare and insurgency during and immediately after this era because of the contentious American involvement in Vietnam (Palka 2000). However, geographer Robert McColl (1969) sought to fill a void in this area, and his "Insurgent State" model furthered our understanding of insurgency as not only a political but a spatial process. His analysis, focused largely on this rural approach following the Maoist model, confirmed that cities and towns were viewed as targets and objectives after securing control over rural, peripheral areas.

Despite the success of the Chinese Communist Party and the appeal of Mao's theories, the rural-based approach to insurgency has not proved successful in every geographic context. Other movements have attempted to follow this model, but political, economic, and sociocultural realities and differences dictate that strict formulas or prescriptions of how to achieve success in every instance are unlikely. The lack of similar success, particularly in Latin America, prompted a transition to urban-based strategies to challenge and topple incumbent regimes (Jenkins 1971; Marks 2003).

In these approaches, insurgent movements sought to attack the ruling authorities' base of power, demonstrating the government's inability to control and secure their own cities and populations, prompting harsh and often indiscriminate responses, and further erode the government's legitimacy in the eyes of the people. However, despite the efforts of charismatic leaders and dedicated followers, most urban-based insurgencies were largely unsuccessful in their efforts to gain political power (Jenkins 1971; Joes 2007). Although Che Guevara's (1985) rural "foco" model became tremendously popular following the Cuban Revolution and publication of his seminal treaty on guerrilla warfare, most theoretical and practical works that emerged from Latin American insurgencies of the period were focused on methods and techniques of urban-based strategies to affect political change in states (Marighella 1972; Hodges 1973). Purely urban-based insurgencies, though, have proved relatively rare occurrences, and research during the 1960s and early 1970s tended to view urban approaches as a component of or complement to efforts in rural areas. As O'Neill (2005, p.162) noted, "For some Latin American theorists like Marighella and Abraham Guillen, action in the cities is crucial but not decisive because the struggle must eventually be transferred to the countryside." A Special Operations Research Office study published in 1965 poses four models of urban guerrilla warfare:

1. Urban demonstration
2. Direct power seizure via urban operations
3. Precipitation of nationwide insurgency
4. Support of rural operations

The first two are distinctly urban-based insurgent efforts and, together with the third model, indicate that the primary effort and focus is in urban areas. In the fourth, however, urban-based insurgent efforts are viewed as secondary, supporting efforts to the insurgents' main effort in the rural areas of the state. Other scholars, though, have contended that even in strictly rural-based movements, insurgencies are inherently urban as the cities, with the dense concentration of population, wealth, and power, and often deplorable conditions for large segments of the population, are the sites where the vast majority of such movements emerge. It is the cities, then, from which the ideas and organization to foment an insurgency evolve and then spread or diffuse to rural areas (O'Sullivan 1991).

These "combined" approaches therefore constitute a third insurgent strategy, with separate but complementary rural and urban components, in which Ho Chi Minh's struggle against the South Vietnamese government and the USA is viewed as the most noteworthy example (Marks 2003). Under this approach, the insurgent movement employs different strategies and methods in different settings to challenge the ruling or incumbent authority.

In assessing the situation in Vietnam (and comparing it to Malaya), Thompson (1966) distinguished the types of insurgent activity within these different spatial settings. He categorized territory under insurgent threat into three distinct areas: populated areas under government control, rural populated areas under disputed control, and remoter lightly populated areas under insurgent control. Rather than categorize by scale or level of settlement (cities, towns, villages), he used the rather ambiguous term "populated area" and noted that each area, generally, was the site of different political and military insurgent activities. In the first, insurgent activity was predominantly subversion, propaganda, sabotage, and minor terrorist acts. In the second, these activities were also evident, but experienced guerrilla activity by small units. The third area saw these activities as well as the operations of larger, more formal insurgent units which acted overtly and exercised effective control of territory.

Other scholars of the Insurgent Era, in analyzing insurgency as a process, subsequently theorized these types of activities as categories as well as stages through which an insurgency progressed. McCuen (1966), expanding the three stages of protracted war, delineated four categories or stages of communist-inspired insurgency: organization, terrorism, guerrilla warfare, and mobile warfare. In analyzing guerrilla warfare specifically in urban areas, Jenkins (1971) theorized five stages: the violent propaganda stage, the organizational growth stage, the guerrilla offensive, mobilization of the masses, and the urban uprising. Ashworth (1991), with a geographer's focus and drawing upon Moss's (1971) work on urban guerrilla warfare, labeled these four stages as preparation and propaganda, urban terrorism, guerrilla warfare, and open insurgency. Ashworth applied these stages toward an understanding of the range of activities and stages of development in an insurgent city. Today, guerrilla warfare receives much less scholarly and media discussion than in previous eras, while terrorism is often treated as a separate and distinct form of political violence from insurgency. Yet the political objectives are similar, and some scholars distinguish the differences between terrorism, guerrilla warfare, and mobile or conventional war as a matter of scale – the size of the group and degree of openness with which they operate (Lynn 2005).

As these theorizations note, insurgents will use a range of violent and non-violent methods to resist or challenge incumbent authority or rule. The particular method, of course, will vary greatly by spatial setting or context, primarily the degree of population settlement and density as well as government presence and control. And as geographers researching war and conflict note, violence and insurgent activity are not evenly distributed or evident across a state's territory, but will be concentrated in time and space, with certain areas experiencing much greater insurgent activity than others (O'Loughlin 2003). Efforts to map the conflicts in Iraq and Syria (particularly in the cities) lend further credibility to Ashworth's conception that different parts of a city will experience different types of violence or insurgent activity, and the result is a patch work of contested areas or neighborhoods (examples of such maps can be found in a host of news media sources such as the *New York Times* or *BBC*, as well as from research organizations such as the Institute for the Study of War).

In most conceptions of insurgency and counterinsurgency, the population is considered the center of gravity (Galula 1964; US Army 2006). But the population cannot be considered separate from the territorial component of where the population resides. As O'Sullivan and Miller (1983) found, "Territory can best be controlled politically and economically by holding its urban foci. Towns then are the primary targets of war." They further predicted that as urbanization levels increased, "any conflict involving or seeking the interest of the mass of the population is bound to take place in the cities." Subsequent works focusing on war and the city further predicted that cities, particularly in insurgencies, would emerge not only as the target or objectives in such conflicts but as the primary sites of violence and contesting state power relationships (Ashworth 1991; Taw and Hoffman 1994).

Since 9/11 and the USA led "war on terror," this trend toward the urbanization of war and insurgency is becoming apparent, and cities (or rather, urban areas) have received greater scholarly focus to understand the changing nature – and location – of war (Hills 2004; Graham 2004; Kilcullen 2013). More recent research argues that the distinction between cities as targets or as the primary sites of politically motivated violence is no longer valid. Cities, as the locations of political, economic, and social power and life in modern societies, are both the ultimate prize in a struggle for power and also the scenes where such struggles will be acted out (Kilcullen 2013). However, clearly not all recent and ongoing conflicts are primarily centered in cities, and the use of "urban places" or "spaces" in lieu of "city" suggests that population settlements of all scales are the key locations and sites of political conflict (Graham 2004; Fregonese 2009). However, in predominantly urban societies, such as Iraq (69% urban), Syria (58% urban), and Libya (79% urban) (CIA World

Factbook 2017), trends suggest that the vast majority of conflicts will be waged in and around the major urban centers, particularly in cases of states with one large primate city. This, of course, has tremendous implications – not only for the insurgents and those countering their efforts but for the civilian populations of these afflicted cities whose lives and livelihoods will suffer greatly.

The City as an Insurgent Operating Environment

While terrain analysis in support of military operations in rural, sparsely populated areas tend to focus on the physical landscape (to include landforms and topography, soils, hydrography, climate, weather), that of urban areas focuses much more on the cultural landscape, analyzing the layout and spatial arrangement of street patterns, building types, heights, construction materials, residential versus industrial areas, and underground structures like sewers, cellars, and underground rail or transit lines (Hills 2004).

Much of the research on the intersection between war and the city, particularly within military doctrine and the field of military geography, has focused to a large degree on the influence and impact of these physical structures on military operations within urban settings. Cities, and densely populated areas with buildings and associated infrastructure, have long presented unique and often imposing challenges for military forces, requiring specialized doctrine, tactics, techniques, and procedures (O'Sullivan 1991; Ashworth 1991; Hills 2004). Similarly, insurgents "must plan for and conduct operations" in a range of built up or urban areas, including urban housing areas, suburbs, industrial areas, and "clusters of buildings in a variety of settings" (Newman 1997).

In conventional armed conflicts, cities have largely been viewed as places to avoid while securing control over these areas through strategies similar to Mao's rural model – to encircle, isolate, and cut off supplies and reinforcements in order to avoid the casualties and costs of urban combat. Most military forces have traditionally viewed cities as constricted terrain, limiting movement, channelizing forces, and reducing the range of weapons systems to create a challenging military operating environment, which require large numbers of soldiers and resources to seize and secure these locations (Dewar 1992; Peters 1996). These attributes have led to conclusions that cities ultimately, but not in every case, favor defensive rather than offensive operations. Initial urban-based insurgency efforts in China, Vietnam, and South American examples similarly suggested that cities favored counterinsurgent (defensive) rather than insurgent (offensive) operations. Urban-based insurgencies were concluded to be at a distinct disadvantage because of the concentration of government forces, large populations in which government sympathizers, agents, and informants could report on insurgent activity, and the speed and ability of police and military to rapidly uncover, target, and arrest or target nascent movements (Jenkins 1971; Joes 2007).

However, other analyses offer contradictory evidence and theorize that cities "provide a physical environment that favors insurgent operations" because of these same characteristics, providing insurgent movement, flexibility, short distances, varied terrain and settings, and concealment among the civilian population (Ashworth 1991). This perspective also suggests that for the authorities attempting to counter the insurgency, cities offer the same constraining environment posed to conventional offensive operations, negating their advantages in firepower, mobility, and numbers.

Generalizations about the advantages or disadvantages for insurgents in urban areas are far more nuanced and context specific than these generalizations suggest. Clearly the built environment of cities and population settlements will have an impact and potentially enable or inhibit insurgencies (and counterinsurgency efforts) (Hills 2004). Indeed, these realizations have witnessed a much greater effort in both research and operational planning to gain a more complete understanding of the specific dynamics of individual urban areas, including but not limited to the physical characteristics of the cities themselves. Rather, many of the discussions or analyses of whether urban environments favor either side during an insurgency focus more on the human landscape and level of technology rather than the physical or built features of the urban landscape. Although many of the works considered so influential during and immediately after the Insurgent Era have greatly shaped the way we think about and conceptualize insurgency as a political and spatial process, these works tended to consider cities and urban areas as almost static, homogenous settings.

Today, as the urbanization of insurgency continues, spatial and temporal scale are increasingly important and relevant concepts, and research and operational planning both require greater resolution of these data (O'Loughlin and Raleigh 2008). Cities, and indeed population settlements of all scales, can no longer be viewed as monolithic spaces that are simply targeted as objectives or environments in which to fight. Rather, urban settlements must be understood and approached as a whole comprised of dynamic and diverse segments, which vary greatly in their social, cultural, economic, and political aspects (Harris et al. 2014). And while the structure, arrangement, and other physical attributes of these settings will continue to be important attributes, the human aspects of these environments are increasingly recognized as the factors which will ultimately determine the success or failure of insurgencies. As insurgent movements evolve and take shape in urban areas, it is the characteristics of particular neighborhoods and areas within the cities that are more likely to determine whether these settings are advantageous or not (Kilcullen 2013).

As urban areas continue to grow both in population and territorial size, cities are becoming ever more diverse in their demographics, particularly when growth is largely through migration. And though cities are viewed as the centers of wealth, power, and opportunity, not all segments of the populations or portions of cities enjoy the same privilege. Conditions for some groups, which tend to be spatially concentrated, may fail to improve or even worsen (in reality or perception), and this fosters resentment which in turn provides the impetus for insurgencies in the first place (Ashworth 1991). In many respects, the growing hardships and deplorable conditions for many modern urban residents, particularly in the developing world, are precisely the factors which spur the rise of such movements.

Furthermore, this growing diversity is evident in the insurgent movements as well. Recent analyses of the insurgency in Iraq, as well as Syria and others around the world, reveal that the forces contesting the incumbent or ruling authority are not homogenous, cohesive movements (Adnan and Reese 2014). Instead, the threat to power is contested among and between varying groups, with different power bases and objectives. Additionally, distinct insurgent groups may strive to achieve these results through various methods. As a result, insurgencies today are seemingly complex endeavors in which different groups wage their struggles in different ways in different areas. While this would seemingly make a cohesive, concerted counterinsurgency strategy easier to formulate and enact, this same diversity in ends, means, and locations within and between urban areas among the different segments of the population makes such efforts extremely difficult to analyze and address.

While the human landscape will play a vital role in the initial and ultimate success (or failure) of an insurgency, the level of technology of the ruling authority will also play a significant role. Electronic and aerial surveillance coupled with well-developed human infrastructure and security organizations severely restrict the ability of insurgent movements to go beyond the initial organization or terrorist phases. However, states which lack these technological and organizational capabilities provide a much more conducive setting for budding insurgent movements to expand their base of support and progress beyond the initial stages of evolution.

Conclusion

The urbanization of insurgency, though, cannot be considered solely in relation to the concentration (and increase) of urbanization levels and rates. In many ways, the current trends reflect greater analysis of operational trends and learning (or passing on) lessons from previous experiences in other conflicts and settings. The lessons of Iraq and Afghanistan against the might of the American military and its coalition partners, many of which were adopted (and adapted) from other insurgent movements around the world, have been and are rapidly disseminated through the speed and availability of the World Wide Web. Indeed, the lessons of fighting the most powerful military in the world have offered numerous lessons and examples for other insurgent groups seeking to contest better organized, equipped, and manned forces. While Afghanistan's physical and human landscapes provide US and coalition forces a distinct set of geographic challenges, in predominately urban societies, insurgent movements are likely to look to the example of Iraq, in which the vast majority of the insurgent activity against the post-Saddam Iraqi Government and the USA – as well as the civilian population – was conducted within the major urban centers. In this way, the often smaller, less well-armed, and supplied insurgents used the urban landscape to offset their tactical disadvantages. However, as the recent struggles to reclaim territorial control from ISIS, mainly in the cities and towns of

Iraq and Syria indicates, these advantages are only offset in urban areas if one (or both) sides operate under strict rules of engagement, limiting (or attempting to) civilian casualties and collateral damage to vital infrastructure.

Furthermore, in today's information age where news is reported rapidly and continuously, urban settings offer much greater locations for insurgents to make their cause known to world opinion. The transition to and adoption of terrorism in the 1970s and 1980s has been attributed, in many respects, to this end (O'Neill 2005), and the proliferation of news sources and media outlets today continues to suggest that incidents of politically motivated violence will garner far more attention than acts in rural, sparsely populated areas less connected to the world economy (Kilcullen 2013).

Current trends indicate that insurgencies will likely remain the most prevalent form of war in the immediate future and that cities – and urban settlements of all scales – will increasingly become the objective and sites of conflict. While scholarly analysis and news reporting lend strong evidence to support these observations, to truly assess the degree to which these trends are true will require further empirical study. In particular, research on the relationship between insurgencies and cities would benefit from continued (and increased) attention from geographic perspectives. Specifically, research along these lines would benefit significantly from more concerted and academically grounded distinctions between rural and urban areas and what constitutes or is categorized as a city, versus population settlements of other sizes or scale.

References

Adnan, S., & Reese, A. (2014). Beyond the Islamic State: Iraq's Sunni insurgency. Middle East Security Report 24 Institute for the Study of War, Washington, DC. http://www.understandingwar.org/report/beyond-islamic-state-iraqs-sunni-insurgency Accessed 2 Jan. 2017.

Ashworth, G. (1991). *War and the city*. New York: Routledge.

Asprey, R. (1994). *War in the shadows: The guerrilla in history*. New York: Morrow.

Black, J. (2016). *Insurgency and counterinsurgerncy: A global history*. Lanham, MD: Rowen & Littlefield.

CIA. (2017) The World Factbook. https://www.cia.gov/library/publications/the-world-factbook/. Accessed 2 Jan 2017.

Dewar, M. (1992). *War in the streets: The story of urban combat from Calais to Khafji*. London: David & Charles Publishers.

Fregonese, S. (2009). The urbicide of Beirut? Geopolitics and the built environment in the Lebanese civil war (1975–1976). *Political Geography, 28*, 309–318. https://doi.org/10.1016/j.polgeo.2009.07.005.

Galula, D. (1964). *Counter-insurgency warfare: Theory and practice*. New York: Praeger.

Graham, S. (2004). *Cities, war, and terrorism: Towards an urban geopolitics*. London: Blackwell.

Guevara, E. (1985). *Guerrilla warfare*. New York/Lincoln, NE: University of Nebraska Press.

Harbom, L., & Wallensteen, P. (2007). Armed conflict, 1989–2006. *Journal of Peace Research, 44*, 623–634.

Harris, M., Dixon, R., Melin, N., Hendrex, D., Russo, R., & Bailey, M. (2014). Megacities and the US Army: Preparing for a complex and uncertain future. Chief of Staff of the Army Strategic Studies Group. https://www.army.mil/e2/c/downloads/351235.pdf
Hills, A. (2004). *Future war in cities*. New York: Frank Cass Publishers.
Hodges, D. (Ed.). (1973). *Philosophy of the urban guerrilla: The revolutionary writings of Abraham Guillen*. New York: William Morrow and Company.
Jenkins, B. (1971). *The five stages of urban guerrilla warfare: Challenge of the 1970s*. Santa Monica, CA: Rand.
Joes, A. (2007). *Urban guerrilla warfare*. Lexington, KY: University of Kentucky Press.
Kilcullen, D. (2013). *Out of the mountains: The coming age of the urban guerrilla*. New York: Oxford University Press.
Laqueur, W. (1976). *Guerrilla*. Little, Brown and Company, Boston: A historical and critical study.
Lohman, A., & Flint, C. (2010). The geography of insurgency. *Geography Compass, 4*, 1154–1166. https://doi.org/10.1111/j.1749-8198.2010.00361.
Lynn, J. (2005). Patterns of insurgency and counterinsurgency. *Military Review, 85*(4), 22–27.
Mao Tse-Tung (1961). *On guerrilla warfare* (trans: Griffith, S. B.). New York: Praeger Publications.
Mao, T.-T. (1967). *Selected military writings of mao tse-tung*. Peking: Foreign Language Press.
Marighella, C. (1972). *Mini-manual of the Urban Guerrilla*. Montreal: Abraham Guillen Press.
Marks, T. (2003). Urban insurgency. *Small Wars and Insurgencies, 14*, 100–157. https://doi.org/10.1080/09592310410001676925.
McColl, R. (1969). The insurgent state: Territorial bases of revolution. *Annals of the Association of American Geographers, 59*, 613–631.
McCuen, J. (1966). *The art of counter-revolutionary war*. Harrisburg, PA: Stackpole Books.
Moss, R. (1971). *Urban guerrilla warfare*. London: Adelphi Papers No 79 International Institute for Strategic Studies.
Newman, B. (1997). *Guerrillas in the mist*. Boulder, CO: Paladin Press.
O'Loughlin, J. (2003). Spatial analysis in political geography. In J. Agnew, K. Mitchell, & G. Toal (Eds.), *Companion to political geography* (pp. 30–46). Oxford p: Blackwell.
O'Loughlin, J., & Raleigh, C. (2008). Spatial analysis of civil war violence. In K. Cox, M. Low, & J. Robinson (Eds.), *The SAGE handbook of political geography* (pp. 493–507). Los Angeles: SAGE Publications.
O'Neill, B. (2005). *Insurgency and terrorism* (2nd ed.). Washington, DC: Potomac Books.
O'Sullivan, P. (1991). *Terrain and tactics*. Westport, CT: Greenwood Press.
O'Sullivan, P., & Miller, J. (1983). *The geography of warfare*. New York: St Martin's Press.
Oppenhiemer, M. (1969). *The urban guerrilla*. Chicago, IL: Quadrangle Books.
Palka, E. (2000). Military geography: Its revival and prospectus. In G. Gaile & C. Wilmott (Eds.), *Geography in America at the Dawn of the 21st century* (pp. 505–513). London: Oxford University Press.
Peters, R. (1996). Our soldiers, their cities. *Parameters, 26*(1), 43–50.
Sanger, R. (1970). *The insurgent era*. Washington D.C: Potomac Books.
Special Operations Research Office. (1965). *Preliminary survey of insurgency in urban areas*. Washington, D.C.: The American University.
Taw, J., & Hoffman, B. (1994). *The urbanization of insurgency*. Santa Monica, CA: Rand.
Thompson, R. (1966). *Defeating communist insurgency*. St Petersburg, FL: Hailer Publishing.
U.S. Army. (2006). *Counterinsurgency. FM 3-24*. Washington, D.C.: Department of the Army.

Index

P. L. Guth (ed.), *Military Geoscience*, Advances in Military Geosciences,
https://doi.org/10.1007/978-3-030-32173-4

MIX
Papier aus verantwortungsvollen Quellen
Paper from responsible sources
FSC® C105338

If you have any concerns about our products,
you can contact us on
ProductSafety@springernature.com

In case Publisher is established outside the EU,
the EU authorized representative is:
Springer Nature Customer Service Center GmbH
Europaplatz 3, 69115 Heidelberg, Germany

Printed by Libri Plureos GmbH
in Hamburg, Germany